REAL DIRT

AN EX-INDUSTRIAL FARMER'S GUIDE TO SUSTAINABLE EATING

REAL DIRT

AN EX-INDUSTRIAL FARMER'S GUIDE TO SUSTAINABLE EATING

HARRY STODDART

IGUANA

Publisher: Greg Ioannou
Editors: Christa Bedwin, Sheila Wawanash, Meghan Behse
Front cover image: Connor Stoddart
Front cover design: Lauren Ogilvie
Book layout design: Meghan Behse

Library and Archives Canada Cataloguing in Publication

Stoddart, Harry, 1969-, author
    Real dirt : an ex-industrial farmer's guide to sustainable eating / Harry Stoddart.

Includes bibliographical references.
Issued in print and electronic formats.
ISBN 978-1-77180-011-2 (pbk.).--ISBN 978-1-77180-013-6 (epub).--
ISBN 978-1-77180-014-3 (kindle).--ISBN 978-1-77180-012-9 (pdf)

    1. Sustainable agriculture. 2. Natural foods. I. Title.

S494.5.S86S76 2013   630   C2013-905465-0
                                                         C2013-905466-9

This is an original print edition of *Real Dirt*.

REVIEWS

"Harry told me this book would prod a few hornet's nests; I had no idea just how extensive and thoroughly he had prodded! Concerned that the entire world needs a wake-up call, Harry Stoddart shares a most insightful look at agriculture and the resulting food that we all eat.... Those three daily choices can be driven by our conscience or the almighty dollar. The law makers who govern food production and those who enforce them, together with those who educate this vital commodity, could all do a better job if they read Harry's book.... Need we say any more? Read the book. You may be shocked but you won't be disappointed."

— Elwood Quinn, La Ferme Quinn, Rare Breeds Canada.

"Harry Stoddart has a wealth of professional, academic, and practical experience in agriculture, and the blend of that knowledge is skillfully demonstrated in *Real Dirt*.... Stoddart recognizes that this book unearths 'sacred cows' on both sides of the debate as to how agriculture can continue to be sustainable. The author advocates for rational dialogue to foster integration of these systems rather than following an either/or philosophy. This book could be the fuel for that discussion."

— Dr. Frank Ingratta, Retired Deputy Minister of Agriculture, Food and Rural Affairs, Ontario

"*Real Dirt* is a thoughtful and well-researched look at our agriculture and food system. It is brought to life with a unique, well-informed perspective and offers practical solutions that will make each of us think about our actions....*Real Dirt* is a must read for anyone who is actually interested in learning about and discussing how to improve our food system for the long term."

— Rob Hannam, Owner, Synthesis Agri-Food Network

I dedicate this book to future generations,

with the hope that I got enough right to be classed as someone who "did something" to improve your world.

TABLE OF CONTENTS

INTRODUCTION

When planning for a year, plant grain. When planning for a decade, plant trees.

When planning for life, train and educate people.

— *Ancient Chinese Proverb*

Our border collie, Kasne,[1] has figured out the difference between chickens and ducks. Our egg layers truly range free, and some of them enjoy coming around the house. Kasne is supposed to keep the hens off the porch, but he generally only does so when someone is looking. He's figured out that it doesn't matter how many times you tell a chicken to get lost, they're all cocky enough to come back. Ducks, on the other hand, are Kasne's favourite livestock.

Our paddling[2] of laying ducks is Khaki Campbell. They are much more wary and much stronger flockers than chickens or any other animal we have. With sheep and cattle, there is a clear lead animal — the alpha female — followed closely by the alpha male if there is a threat. With ducks, there is no clear leader, but they move with the precision of an Olympic synchronized swimming team.

The Khakis fascinate Kasne. He will spend hours at the pond, watching and manipulating the movement of the ducks. He'll take a step to the left and they will all simultaneously swim to the right. He spent last winter subtly moving them around the yard and over snowdrifts. When spring came, the ice was slow to come out of the pond. For a couple of weeks, there was thin ice or open water around the edge and a

---

1. It's pronounced KAHZnee. He arrived with the name from a border collie rescue, and I don't know what it means.
2. Several collective nouns can be used for ducks, including flock, raft, team, brace, flush, and paddling. I like the term "paddling" because it so suits our ducks, and besides, we already have a flock of sheep and a flock of chickens. Some days we even have a murder of crows around.

piece of thick ice floating in the middle. The ducks learned that Kasne wouldn't follow them out onto the pond — the cold ring of water around the outside stopped him in his tracks — so the ducks would calmly sit in the sun and paddle in the puddles on the ice, safe in the knowledge that the dog wouldn't come out on it. As the spring progressed, the ice gradually melted until the pond was completely open.

One day, I was sitting in my office overlooking the pond with the windows open to the warm spring breeze. I heard a splash, followed by a lot of cursing and swearing from the ducks. Kasne had splashed into the pond and was swimming after them while they paddled and flapped their wings as fast as possible to get away. Apparently, it hadn't occurred to them that the dog could swim. In their relative experience, a dog had never ventured into the pond. There was no organized retreat; they all panicked and went off in every direction possible.

What Kasne and the ducks had discovered was the limits of their relative perspectives. But we all come at life with our own perspectives and biases. In fact, everything we experience, we experience relative to our knowledge and accumulated wisdom. Every time you eat, for example, you make a choice about the type of food you consume. Your choice is based on your relative experience and goals. And your choice signals the market to produce more of what you chose to eat. All of our choices, taken together, are determining which type of food system produces which types of food now and in the future. In essence, you are choosing the food that will be on your grandchildren's buffet. You are also choosing how that food should be produced on the farm. If you choose solely on the basis of price, you are ensuring that lower-cost food will be pursued regardless of the non-monetary costs to society (for example: erosion of land, phosphorus pollution of rivers and lakes, and greenhouse gas emissions). If you value more than just what is produced at what price, your choices need to reflect your other values. But what should we value to ensure that our grandchildren have healthy, abundant food?

This is a question that is not easily answered. I've been farming for over two decades and I have trouble sorting through all the information and misinformation about food production. If our primary concern is preserving the ability to produce food abundantly for future generations, what choices should we be making in the present? Should we denounce genetically modified food? Should we choose organic? Should we become vegetarian? Should we be locavores? Are there factors beyond what we can read on labels? How do pesticides, fertilizers, and antibiotics fit into the equation?

Let's get one thing straight off the top — there are no magic labels to guide you. If we want to sustain humanity's ability to harvest energy from the sun and turn it into protein and calories — which is what agriculture is and does — we need to look beyond our conventional systems, beyond organic, and beyond vegetarianism to a system of food production and consumption that is a closed loop. Nutrients, water, organic matter, and fuel are all making a one-way trip through our food system and being discarded. All of our current production and consumption systems have serious flaws. We need to reduce our inputs of energy, nutrients, and water, but we also need to design our systems so that the resources that *are* consumed are returned to be reused. Pumping water out of an underground aquifer to irrigate is folly if the aquifer is not being recharged. Mining phosphorus and ultimately flushing it down the toilet can only continue for so long. We do not have limitless supplies of fossil fuels to use to convert the nitrogen in the air into a plant-usable fertilizer — especially when you consider the fact that there is an entire order of plants[3] that work symbiotically with bacteria to fix nitrogen using only solar power.

The challenge is to stand back and observe the system as a complete system. As I noted above, we all come at life with our own perspectives and biases. We need a critical eye to discover what lies under and supports the patterns we see and experience. In the agriculture and food system, there are a lot of groups yelling at each other and very few trying to understand the other points of view because they are convinced they know what the "truth" is. On one side we have activists attacking a particular issue within agriculture, and on the other side we have a group of farmers who feel their livelihoods and integrity are being threatened. It is tough to have an informed dialogue in that environment. There is wide disagreement even within agriculture on the way forward (thankfully, because none of us has it "right" and "groupthink" can wreak devastation faster than indecision).

First we need to set aside all the current perceptions of what a farmer is and does. We need to reflect on what the ultimate goal of agriculture is from a society's point of view and a farmer's point of view. Society relies on farmers to capture solar energy and turn it into edible energy and protein. Society's preference would be that this is done at the lowest

---

3. There are also a number of nitrogen-fixing organisms scattered across the various kingdoms of biology.

cost possible, and some prefer that their food be provided without diminishing the ability of their descendants to eat. From farmers' perspectives, agriculture is a vocation that they use to create the lifestyle that they and their dependents desire by capturing solar energy and turning it into a saleable good.

The challenge is that society's interests and farmers' interests are not always fully aligned. The citizens of Louisiana bear the cost of the dead zone in the Gulf of Mexico at the mouth of the Mississippi River. The farmer in Illinois who contributed a very small amount of soil and pollution to creating that dead zone bears none of the cost. Eventually enough soil will be lost to cause harm, but in any given year, if the choice is between conserving soil and making a profit, making a profit wins. It's not that the farmers don't care about the environmental impacts of their choices, it's that there is no consequence for ignoring them in the short run.

A farmer is focused on managing four main ecosystem processes: the flow of energy, the cycling of water, the cycling of minerals, and the changing population diversity.[4] I'm going to use the term "ecosystem" often here. It's a term that wasn't coined until the 1970s, even though these systems have existed since the beginning of time. However, I want to stretch the concept of an ecosystem a little. Most of us think about "nature" when we hear the word: the circle of life where a plant grows, is eaten by a herbivore, which is eaten by a carnivore, which dies and is turned into plant food by a number of microorganisms to feed another plant. That is an example of an ecosystem that is visible at "human scale." With the exception of the microorganisms, we can see each of the actors and easily identify most of the role players. We talk about the forest ecosystem, the marine ecosystem, and the prairie ecosystem as if they are all distinct and discrete. If you are flying at 30,000 feet, it looks pretty easy to mark where the ocean ends and land begins. Next time you are walking along a beach, however, I challenge you to put a mark on the ground where the ocean ends. The low tide mark? The boundary of saline water permeating the ground? (If you choose the latter, at what salt

---

4. These concepts are borrowed directly from Allan Savory's Holistic Management. The best starting place to dive deeper into Holistic Management is Allan Savory's book, co-authored with Jody Butterfield, *Holistic Management: A New Framework for Decision Making*. The following websites also provide a number of additional resources on Holistic Management: www.holisticmanagement.org, www.savoryinstitute.com, and achmonline.squarespace.com.

content does water turn from saline to fresh?) The reality is that ecosystems do not have defined boundaries; there is always an intermediate ecosystem in between that has characteristics of both.

There is an additional dimension to ecosystems that we rarely consider. As people, we live in an ecosystem, an ecosystem lives within us, and there is an ecosystem that lives on us. Plus, all the ecosystems on the planet are interdependent. Are you familiar with fractals? Fractals are objects (real or imagined) that have an identical structure regardless of the scale you examine them at. A tree is a fractal structure. If you examine the twigs, branches, and trunk of a tree, you will see that the same basic structure is repeated at an ever-declining scale as you move from trunk to branch to twig. The same is true of a cauliflower or broccoli head. River systems are repeated patterns as you scale up from creeks to brooks to rivers. The ecosystem is the same. The ecosystem of the planet as a whole is an interdependent connection of all the macro ecosystems (tropical rainforest, desert, steppe, savannah, etc.), and each of these ecosystems can be broken down into a smaller subset of interdependent organisms. Every change we make to one ecosystem has the potential to change every other ecosystem.

Agriculture cannot be separated from the planet's ecosystem. While most industries occur within an enclosed space and what they extract from and deposit into the environment is easily measured, if not always visible, they have little dependence on the environment and its life forms for success. Agriculture, despite the industrialization of the twentieth century, is wholly dependent on ecosystem processes. We use life forms to capture solar energy and turn it into edible energy and protein. The primary life forms we use are dependent on an array of other organisms to provide life-giving substances. We are also dependent on an ecosystem to convert the products of agriculture into usable energy. Without a thriving population of flora and fauna in our digestive systems, we wouldn't derive much benefit from most of what we eat.

This is the context within which we must examine the sustainability of any particular agricultural method. And this is the mistake that most activists attacking agriculture make: They focus on a single aspect of agriculture and pronounce one particular crop or production method either good or bad. For example, it is commonly argued that meat should be eliminated from our diets because too much land is wasted growing crops to feed to animals and we would have a lot more calories available

5

for human consumption if we simply eliminated the animals from the food chain. If it were only that simple. There is no properly functioning ecosystem on the planet that doesn't include animals — both herbivores and carnivores. One of the best soil-building and soil-healing methods in agriculture is a well-managed perennial pasture. It's also one of the most resilient production systems. Poorly managed grazing, however, is extremely destructive.

When you look at all the energy inputs involved in the end-to-end process of capturing solar energy and converting it into edible calories and protein for human consumption, would you expect that a soybean field or a well-managed grazed dairy cow would be the most efficient option? The answer is the grazed dairy cow, and that's without adding in the soil-building properties of the dairy cow. I can also show you dairy farms that are incredible wastes of water, land, and oxygen. But when you're standing in the grocery store, you're given very little information about which production system the milk or cheese comes from, and because of this, what you choose probably has little to do with the environmental impacts.

My relative experience started as the son of a small hog farmer using conventional methods — pesticides, animal confinement, fertilizers, etc. I started my farming career by buying my parents' farm. I almost lost it, spent close to a decade working in the financial towers in the city, and gradually reinvented myself as a certified organic farmer who believes organic principles should be considered only the starting point for lessening agriculture's impact on people and the planet.

Using my perspectives on the system as a whole, I will walk you through each of the major issues surrounding food production and consumption. I will look at the issues from the perspective of a conventional farmer and the perspective of an organic farmer. Blended into the discussion is scientific research that has convinced me to take the stands I have in support of sustainability.

"Sustainability" is a word that is thrown around indiscriminately by many groups. Sustainability is a concept that is simultaneously ephemeral and eternal. It is tough to grasp all the nuances and dimensions. In the strictest definition, if a system is sustainable,[5] it can continue indefinitely. I think there are very few sustainable systems in

---

5. This is one place where I depart from the commonly used definition of "sustainable." I don't believe that a product or a thing can be sustainable. The system that produced them can be sustainable, but the object itself is not.

the world today. Conventional agriculture is not sustainable; it is consuming nonrenewable resources — fossil fuels, fossil water, fossil nutrients, and fossil organic matter — at an alarming rate. It is producing fossil food. But organic agriculture as defined by the current set of rules is not a truly sustainable system either. The rules focus on limiting the use of synthetic pesticides, fertilizers, and antibiotics. Other dimensions of sustainability are still missing. For example, recycling post-consumer nutrients (such as phosphorus) is prohibited, and the phosphorus flowing through our sewage treatment plants is a nonrenewable resource and a pollutant. A sustainable system must find a way to recycle phosphorus and the other nutrients that make a one-way trip from the farm to the city. Erosion is another challenge where organic agriculture has trouble claiming superiority to some methods within conventional agriculture.

I used to favour The Brundtland Commission's definition of sustainability:

Sustainability is meeting the needs of the present without diminishing the ability of the future to meet its needs.[6]

Unfortunately, it seems this definition has been reinterpreted by many to mean:

Meeting the needs of the present in less harmful ways.

I now think John Ehrenfeld has truly captured the essence of my understanding of sustainability in one phrase:

Sustainability is the possibility that humans and other life will flourish on Earth forever.[7]

Even if a system has the potential to be sustainable, there are no magic protections from humanity's greed. A colleague uses a birchbark canoe as an example of a sustainable product. I disagree for two reasons. First, you cannot declare a product to be sustainable without knowing the system within which it was produced. Second, just because it is possible that a system producing a birchbark canoe can be sustainable does not mean that any particular canoe was produced sustainably.

The sustainable birchbark canoe is produced from a mature birch tree and other 100% biodegradable materials accessible by human effort. It is

---

6. United Nations, "Our Common Future," Part II.II.I
7. John Ehrenfeld, *Sustainability by Design*, 49.

then used to transport its owner and his goods without anything more than the water cycle and the renewable solar energy that powers the person's diet. For the life of the canoe, another birch tree is allowed to grow so that when the canoe reaches the end of its life, the canoe can be allowed to decompose in the forest and another tree is available from which to harvest a new canoe. This cycle can continue indefinitely until the sun grows large and bakes the earth in about a billion years. However, if the population of the people harvesting birchbark canoes grows large enough that they are harvesting canoes faster than the trees are regenerating in the forest, the system is not sustainable. A birchbark canoe has the possibility of being produced sustainably, but it is not inherently sustainable.

So if conventional agriculture is not sustainable and organic agriculture is not sustainable, are we screwed? Maybe. History has several examples of sophisticated urban societies that outgrew the ability of their agricultural methods to feed them. We even have examples of indigenous hunter-gatherer societies that caused the collapse of their life-sustaining ecosystems and crashed their populations. However, if we examine the research and look to examples of paradigms that successfully conserve and recycle resources, I think there is a path through the end of the fossil fuel era that does not have to involve civil strife. Whether we choose to follow that path is another story. But be assured we can choose. The choices will not be easy and there will be many voices shouting alternative views.

One of the problems we encounter when looking at many of the issues in agriculture is the timeframe over which we are going to conduct analysis. There is a project in the United States called The Long Now. They are building a 10,000-year clock where the "second hand" ticks once a year. Their purpose is to try to draw attention to the fact that most of our planning horizons are far too short to accurately assess the long-term impact on our species and the planet. I struggled to reach definitive conclusions about a number of the issues detailed in this book until I took the 10,000-year perspective. Once I took a truly long-term view, the best course of action was fairly simple to determine. For example, there is disagreement around how long our existing reserves of phosphorus will last. Some argue we have already passed peak phosphorus and will be scrambling by the end of this century. Others argue we have sufficient known reserves to last several centuries and we will discover more before we run out. The difference between the two viewpoints looks irreconcilable until you look back from 10,000 years in the future. Whether we have one or several centuries worth of phosphorus left, or even a couple of millennia's worth, is moot. The pollution from the phosphorus consumed by humans in their food will collapse what is left of our aquatic ecosystems if we don't start recycling phosphorus.

I have farmed on both sides of the tracks. I have existed in both rural society and urban corporate society. I was never a blind idealist. I am a businessman. I have provided policy advice to almost every federal and provincial government in Canada. I have worked alongside the minds at Canada's foremost agricultural think tank. I believe passionately about preserving the environment for future generations. However, don't try to pigeonhole me; you won't succeed. I have voted for five different political parties in the 24 years I have been enfranchised. This book is part memoir, part exposé, and hopefully entertaining. What I hope you'll find is an informative walk through the facts and arguments that gradually took me from someone who switched to organic as a desperation strategy to save a family farm to someone that strongly believes in the principles[8] that founded the organic movement many decades ago but also believes that those principles were just the beginning of the discussion, not the end point.

I've attempted to present a balanced view of the facts and the logic that have led to the conclusions I have reached. I discuss each of the major areas where agriculture is justifiably open to criticism. What you won't find are strident declarations of the superiority of any particular method or technology without a reasoned explanation for why I draw the conclusions I do.

For those who would accuse me of being a Luddite, I bristle at the suggestion. The future of agriculture cannot simply be a return to the methods employed prior to the development of synthetic pesticides and fertilizers. Those methods wore out farmland and forced the migration of entire ethnic groups. Modern-day Turkey is part of what is historically referred to as the Fertile Crescent, a region that saw the domestication of more plant and animal species than any other region on the planet. Yet today, vast portions of the region are largely desert wastelands. More recently, the great push west in the United States was partially a function of a great decline in the productivity of land in the east. Each successive push west was the result of farmers looking for fresh land that was more productive to farm. From the Hahn Dynasty to the Roman Empire to the

---

8. These principles do not necessarily include all of the rules that currently define organic.

Inca Empire, the history of civilization is littered with impressive agricultural systems that fed vast urban centres but that ultimately failed. Our vision for agriculture has to incorporate all the knowledge we have and move to an integrated system that approaches sustainability, lest we repeat the mistakes of our forefathers.

But neither am I blindly accepting the technologies that modern industrial agriculture embraces. Modern agriculture cannot continue on its current path. It consumes too many fossil resources — oil, water, organic matter, soil, and fertilizer — to be sustainable long term. The question is not whether we should change our methods, but how are we going to change to sustain food production into the future?

To make informed choices, we need to cut through the haze of empty rhetoric and understand the real issues. Most of you will recognize the scientific term we use for empty rhetoric down on the farm — bullshit. Pardon the vulgarity, but there is no other way to accurately convey my feelings about a lot of the information that flows around our consumption of food. The labels on processed food are bullshit. They represent the closest a company can get to outright lying about their products without false advertising laws or even tort law kicking in. Disagree? Froot Loops "is a fun part of a complete breakfast, and is a good source of fibre." Coca-Cola's Full Throttle energy drink is a "natural health product," and "whole wheat" bread does not contain the whole grain.

Most of the discussion of the differences between organic and conventional food production is bullshit. Organic tries to portray conventional production as raping and polluting the environment. Conventional tries to portray organic as a return to a nonexistent idyllic past that will cause the starvation of millions. Neither view is accurate. The truth is, neither system is sustainable. We need to move beyond both to survive more than another century with current levels of calorie and protein production.

A good portion of the discussion around genetic modification (GM) is guided by misinformed bullshit. The industry paints it as benign technology that will save the world, and the opponents paint it as unleashing the four horsemen of the apocalypse. Neither view is accurate, or helpful. There are life-affirming implementations of GM technology, but there are uses that do not pass a simple cost–benefit analysis.

Any discussions of pesticides invariably turn to residue levels and the toxin load in your food. Two arguments are often repeated and widely believed. The first is that all forms of food before the industrialization of agriculture were benign and free from toxins and we can return to that idyllic form of production simply by banning all use of pesticides. The second is the notion that the current biologically based, dose-response science of toxicology represents a good framework for establishing the safety of pesticide and toxin residues in our food system and a concomitant regulatory framework.

If your biggest concern about the use of antibiotics in food animal production is that you might ingest some antibiotic residues in the meat you consume, you have swallowed bullshit. The industry would be happy if that is where your thinking stopped, because they are able to produce study after study that shows you do not have to worry about antibiotic residues in your meat. Somewhere between 70% and 80% of antibiotics used in the United States are fed to animals. They are fed at levels that Alexander Fleming, the discoverer of penicillin, warned in his Nobel Laureate address in 1945 would lead to the creation of antibiotic-resistant bacteria.[9]

Do I have your attention yet? I started out as a conventional farmer after studying at a globally recognized university, earning both a bachelor of science in agriculture and a master of science in agricultural economics. I confess, I have planted genetically modified organisms (GMOs), used synthetic pesticides and fertilizers, caused unnecessary erosion, and fed subtherapeutic doses of antibiotics to pigs. I embraced all forms of technology to make my operation more efficient. I grew GMO corn the very first year it was on the market. I increased our farm's use of Roundup significantly as I started converting to no-till farming.[10] I renovated our barns to use liquid manure rather than solid so that I could run double the number of pigs that my parents had with the same labour. I have farmed almost 2000 acres both conventionally and certified organically. I have raised livestock in a confinement operation and in an extensive pasture-based system. I followed the path and analysis taught by the professors under whom I had studied. I had the same conscience then that I do now. Yet today, a little over 15 years later, I am a grass farmer raising grass-fed beef and lamb as well as pastured poultry and pork on certified organic land that may never see another tillage machine disturb the soil.

---

9. Alexander Fleming, "Penicillin."
10. Roundup is the most widely used herbicide globally. The active ingredient in Roundup is glyphosate.

What follows in this book are the views of a farmer who has farmed in both worlds, understands both the warts and benefits of both systems, and has made choices to move his farm closer to a truly sustainable operation than either system prescribes on its own. Follow my journey as I describe a reformation that started with purchasing my parents' farm. I'll give you insight into the choices I made, and the motivation for them. Not everything I have done has been motivated by ecological sustainability — some of the decisions were an overt attempt to survive economically. Not every decision has been an optimal choice for either long-term economic or ecological sustainability. I describe everything, from the heartbreak I felt when I faced bankruptcy to the hope I feel as I build a farming and marketing model that will allow me to look future generations of Stoddarts squarely in the eye and genuinely believe that I have left them a piece of ground that is more productive and healthier than the one that I started with — that I have done what I could to make their world more sustainable than the one I lived in. I've stepped in a lot of bullshit along the way; hopefully I can help you avoid swallowing any more than you absolutely have to.

The quickest transformation to sustainable methods will not occur through legislation. The greatest lever we can pull to shift agriculture is the power of the consumer. I have written this book to educate eaters so that you can make informed choices. Through your informed choices, the market will respond significantly faster than any legislative process would. An informed eater can accomplish what no politician would dare attempt. The collective action of eaters is the only force large enough to meet the significant threats we are facing head on and succeed.

From my perspective, the greatest threats to humanity's ability to feed itself are not currently on the radar of most eaters. It is not the expanding population. It is not the differences between certified organic and conventional production. It is not the irretrievable release of biotechnology into the environment. It is not pesticides and their environmental impacts. These are all important issues that need to be discussed and ameliorative strategies developed. But there are three agriculture-related issues that, left unchecked, will result in societal strife in Western democracies within our lifetimes: antibiotic resistance, erosion, and climate disruption.

There are solutions to these problems. We can alter the course of history with existing technology and knowledge. The good news is that we can have it all — healthier people, a healthier environment, and an adequate supply of protein and calories to meet the needs of our expected peak population. We as citizens can accomplish this without the intervention of governments (actually, their reduced intervention would go a long way). The final chapters set out a set of guiding principles to help you change your choices in favour of true sustainability.

Change will require some completely new paradigms. Some of the answers run against conventional wisdom. I'm partway to achieving the vision, but I still have many steps to take. In a previous subtitle for the book, I referred to myself as a "reforming" industrial farmer. It may strike many of you as an odd turn of phrase, but it applies because of the journey that I have been on and my recognition that the journey is far from over. I have planted grain. I have planted trees. Hopefully, this book will plant information for some — inform you in a way that guides you to choose sustainable food production for today and the future.

CHAPTER 1

DESTINY'S CHOICE OR CHOICE OF DESTINY?

If a product is more expensive than another one, and more sustainable in ecology, consumers will not buy it.

— *Jean Claude*

Our dogs are an integral part of our farm security system. We have two. You met Kasne, our border collie, in the introduction. Our second dog, Lilly, is a Maremma–Great Pyrenees cross. Border collies are classed as stock dogs (herding dogs) and Maremmas and Great Pyrenees are classed as guardian dogs. They don't herd animals, but they'll chase off anything that doesn't belong. If the intruder doesn't heed the warnings, they will attack. Viciously.

Lilly is the most observant beast I have ever met. It's her job to identify threats early and scare them off. However, she defines anything that is different as a threat. The good news is that a coyote trying to sneak in for a free meal has little chance of success — she sees (or smells or hears; I'm not sure which) them coming at a long distance and starts sounding the alarm and heads out to chase the intruder off.

The bad news is that she hits a few too many false positives. When a bright red full moon appears on the horizon, she sounds the alarm. When hunting season starts, she sounds the alarm. The worst incident was once when my in-laws, who live in a second house adjacent to ours on the farm, had been away for several months. The first night they were home, there was a light on in the house, and Lilly let the whole neighbourhood know that something was different. Do you remember the "twilight bark" from the original *One Hundred and One Dalmatians*? That was how our quiet little corner of Mariposa Township sounded that night. Every dog within dog earshot was relaying the message — all because Lilly noticed a light on in the in-law's house that hadn't been on for months. We have Lilly so we can sleep better at night, knowing that our livestock are protected. That night, no one in the neighbourhood slept well.

I see a lot of Lillys when I look at our agriculture and food system. Every other day, someone is sounding the alarm about some aspect of what we eat. Some of these alarms are justified, but some of them make me shake my head. I recently saw a picture of a GMO protester holding a sign that read, "I don't want DNA in my food." So your diet is what? Coke and cream soda?

With Lilly, it's easy to figure out whether she is sounding a false alarm. First you listen for Kasne. If Kasne is still snoozing on the porch, then there isn't much to worry about. If the two of them have run off barking, however, you know there is a real threat and you listen to how close or far away they are. If they're in close, you probably need to grab a firearm and provide backup.

If we extend the analogy to agriculture, instead of Kasne confirming Lilly's alarm, we would have Kasne and Lilly arguing about whether the alarm was real while the coyotes snuck in and helped themselves. Agriculture has many competing views on how resources should be managed as well as many competing views on which practices should be considered sustainable and which practices should be sounding alarms.

When you strip away all the externals — crops, livestock, technology — agriculture is an industry that captures solar energy and transforms it into energy (calories), protein, fibre, and leather for human consumption. Some farms, for example vegetable farms, perform the complete transformation, whereas others, for example hog farms, provide only one step.

It is this inherent reliance on solar energy that gives agriculture the possibility of being truly sustainable. The carbon, oxygen, nitrogen, and hydrogen that agriculture assembles into food are freely available from the atmosphere and the remainder of the minerals required could be 100% recycled. All the minerals that we consume in food can be returned to the soil and reused in a never-ending loop.

Modern agriculture, however, and even historical agriculture, does not operate this way. Modern agriculture is plagued by unsustainable uses of fossil fuels, fossil water, fossil organic matter, and fossil nutrients. Today's agriculture is much closer to an extractive industry than the naturally recycling system that it could be. Like the birchbark canoe, our food supply has the possibility of being sustainable, but the current system providing that food supply is not. Most of the inputs make a one-way trip from mine or well to farm to city.

But of course agriculture is also and always will be a business. The journey of our own farm from a farrow-to-finish (birth-to-slaughter) confinement hog farm to a grass-fed beef and lamb farm, for example, has been punctuated by a few clear market signals.

We purchased the pigs and machinery from my father and mother in 1995. Pork prices were strong, crop yields were decent, and we had a good first year. In February of 1996 our second son was born. In July I put together a plan to buy the home place from Mom and Dad and simultaneously double our sow herd. We took possession on October 11 and poured the last of the concrete for the expansion on November 18, covering it with straw to insulate it and allow it to cure properly before freezing.

Ten months later I was sitting in our bank manager's office renegotiating our financing. The most devastating thing that could have happened (short of one of us getting killed) had happened. Our supplier of gilts (female pigs that haven't had a litter yet) was negative for porcine respiratory and reproductive syndrome (PRRS) when we signed the deal to buy our expansion stock. By the time he was delivering the animals, they were infected with PRRS. I'm not sure at what point he knew, but he never told us. I would have stopped taking delivery and immediately started vaccinating if I had known.

But I didn't. The first indication we had that something was wrong was when our near-term abortion rate sharply increased. Sows that were within a week of their due dates would abort. We got to count how many piglets would have been born, but instead of watching them grow, we shovelled up the dead bodies. Every morning when I woke up, one of the first thoughts that went through my head was whether I would find another aborted litter that morning. It was one of the toughest things I have ever lived through. It turned out that it was also one of the luckiest.

Because of the PRRS outbreak, we marketed fewer pigs in the year after the expansion than the year before. We ran out of cash in August and had to renegotiate our financing. The bank agreed but margined us, meaning that we had to prepare and submit a set of financial statements every month. Through the fall of '97 our financial position gradually improved as the impact of containing the PRRS outbreak took effect and our inventory of growing pigs increased. At the same time, the price of pork was gradually declining. By January it was well below our cost of production. When we prepared our statements at the end of February, we had two consecutive months of decline in our financial position. The futures prices indicated we were in for several more months of the same.

I lost one night of sleep. I got up in the middle of the night and went out to do chores. By the time Silvia came out at around five o'clock in the morning, I had the bulk of chores done and she found me sitting on the floor in one of the sow barns, thinking. Being an agricultural economist, I could read financial statements about as well as most bankers. I had crunched the numbers in my head every way I could think of and had come to the inescapable conclusion that we would be better off if we shut down the hog operation and I got a job. Did I mention that we had found out in January that Silvia was expecting our third child? By three o'clock that afternoon we had informed the bank and my parents.

Then I had to deal with the shame I was feeling. I was smart enough that this should not have happened. I had graduated at the top of my high school and undergraduate classes. I was one of only a handful of farmers in our county with a master's degree. When I came home to farm, one neighbour even asked me straight to my face if I was going to show everybody how to farm. My parents had made a heavy bet with their retirement savings that I would succeed. I resigned from the board of the county pork producers' association by letter — I could not face actually walking into the meeting.

I had put together an aggressive plan, but that is what we needed to carry the mortgage to purchase the home place. I had farmed for 18 months, living in a rented house a couple of miles from the home place. When a thunderstorm went through in the middle of the night, I would get up and drive to the home place to check on the barns because the two places were on a different electrical grid and there was no way of telling whether the hydro was out. If it was, I needed to manually open a number of doors to prevent the pigs from suffocating in the barns while the fans were off. Mom was doing the bedtime barn check so I didn't have to drive back to check for any problems before going to bed. (It's amazing what pigs can accomplish in a few hours — the main reason was to check that temperatures were OK and that nobody was farrowing.) To do the best job managing the hog operation, we needed to be living at the home place.

It had taken me a while to figure out a financial plan we could carry. It took less time for it to fall apart. It was hard to look people in the eye when I was projecting my judgment of myself onto them. Some of them *were* judging me. Most were not. However, I didn't have enough life experience at that point to understand that most of the successful farms I saw around me had had their own close scrapes with financial disaster and most farmers knew that they had been lucky to avoid ending up where I was. When you first start out, you are skating pretty close to the edge and it doesn't take much to push you over.

A series of market signals arrived shortly after we made the decision to exit hogs in Q1 1998. Hog prices averaged just over $135 per 100 kilograms (ckg) in Q3 1998. They plummeted in Q4 to an average below $85/ckg. Our cost of production at that time hovered around $135/ckg. Since we were already in the process of exiting the hog business, the market crash didn't have a huge impact on us. However, it provided confirmation that we had made the correct decision. There is no way we would have survived, and in a rather ironic twist, the devastation of the PRRS outbreak had saved us from bankruptcy. Barely.

The second signal came from a neighbour who explained the economics of organic production to me. We grew our first certified organic crop in 1998 and those margins helped us crawl out from under our financial predicament.

The third signal came from within the organic grain industry. In the early 2000s it was becoming clear that the organic grain industry was transforming into a global commodity market. We were loading shipping containers of spelt to float to Europe. We had to hold beans longer than expected one year because Brazil was undercutting North America in the Japanese soybean market. The watershed event that started us thinking about moving away from organic grains and oilseeds came from an unexpected competitor in the organic flax market. We were growing organic flax, cleaning it to a food grade specification and marketing it in the Greater Toronto Area, primarily to bakers and suppliers of health food stores. Demand was greater than our supply and we were trying to purchase additional flax to meet our buyers' needs. We called everyone we knew across Ontario and Western Canada. Unfortunately, it had been a poor crop year in the West and there was no organic flax to be found.

We eventually found some. We could have had a 20-tonne container load delivered to the farm. The flax came with a full set of certified organic papers and was already cleaned to a food grade spec; all we would have to do is package it. The price was less than what I had been paying to neighbouring farmers straight off the combine. It would have been a good deal for us, but it was a concerning one for our future in the organic flax business.

The country of origin was China. We could buy fully cleaned flax from China, delivered to our farm, for less than the price of flax straight off the combine in Ontario that still needed to be cleaned. We decided to post a "sold out" sign for flax for the balance of the season. But the message was clear. The organic grain business was heading in a direction that looked a lot like the conventional agricultural commodities businesses that we had so recently exited. Competing in the global grain market meant we would have to continue to increase efficiency and lower costs to match our lowest-cost competitors. We would be forced to match the corners they were cutting and focus on meeting the letter rather than the spirit of the organic regulations.

We decided we needed a direct-to-consumer business in which our customers valued what we were doing and would not sell us down the river for a few dollars a tonne. It took us five years to settle on our current style of Community Supported Agriculture (CSA) meat share program. Consumers purchase a "share" of our meat harvest for six months. We get the commitment in advance and all the meat that leaves the farm is sold before it goes out the lane. The CSA business model allows us to form a committed relationship with our customers that we feel is sustainable for the next number of years if we can find enough consumers who understand the relative environmental sustainability of our practices and are willing to pay a price that sustains us economically. Is this the place for us to stop and admire where we are? No. To use a hockey metaphor, you always have to keep your feet moving. As soon as your feet stop moving, you are vulnerable.

At the Stoddart Family Farm, we've found an economic model that we believe is sustainable, but how do we ensure that our decisions in all the other dimensions of farm management lead to sustainable outcomes? We are telling the marketplace that we value sustainability, but there is no official definition of this term. Following certified organic principles gets us partway there, but as you will discover in subsequent chapters, I have concerns with organic's approach to a number of issues. If you look at what nature produces annually in ecosystems around the world without any interference from man, in spite of interference from man, it is truly amazing and has been sustained for tens of thousands of years. Agriculture isn't 10,000 years old yet, and it has developed an incredible knack for destroying productive soil and creating deserts where ecosystems had previously thrived.

Many people equate "small" with sustainable. I believe that sustainability is a concept that is largely independent of farm size. True sustainability presents different challenges at different scales. To generalize, I would posit that economic sustainability is the greater challenge in small-scale agriculture and some dimensions of environmental sustainability can become more challenging in large-scale agriculture. For a small farm to sustain a farmer and provide a decent living, that farm needs to be dedicated to producing high-value crops and capturing as high a percentage of the retail dollar as possible.

The environmental challenge on larger farms is most obvious in livestock. When you have a large concentration of animals in one spot with feed being hauled to them, the manure becomes uneconomical to haul sufficient distances to prevent the buildup of nutrients on the land around the operation. Human ingenuity has turned an extremely beneficial natural nutrient recycling process into two problems: pollution at the livestock farm and nutrient deficiency on the crop farm. But this problem also occurs on small mixed farms. The mineral levels in fields closest to the barn will invariably be higher than the levels in the fields farthest from the barn. The mindset of the manager and economic rewards have more influence on sustainability than the scale of a particular farm.

Many point to "industrial" production models and the evils of capitalism as the source of many of the problems in modern agriculture. Public policy and agriculture are so intertwined, however, that it is impossible to accurately assess what results from the natural forces of the market and what is the result of government interference in agricultural markets. "A chicken in every pot" is the quintessential statement of the importance of a well-fed population to governments. The expression was most recently used in advertisements by the Hoover campaign in 1928.

Government agricultural policy and interference in agricultural markets has distorted prices for most agricultural commodities for much of the last century. Even the current prosperity on the back concessions has a direct tie to US government ethanol policy. The US government provided subsidies to relaunch the ethanol industry when the price of ethanol was higher than its energy equivalent from petroleum sources. Without the subsidy, only a fraction of the investment in ethanol would have occurred. Even today, the minimum renewable fuels standards are driving an artificial demand for ethanol. Because the US standard includes a requirement for ethanol from advanced sources (i.e., not corn), there are currently tankers hauling ethanol from Brazil (derived from sugar cane, which is considered an advanced source) to the US and tankers hauling US corn ethanol to Brazil (because the corn ethanol is priced lower than the sugar ethanol). The bureaucrats' intentions were good (but you know what they say about good intentions). They intended to jumpstart the renewable fuels industry to reduce dependence on fossil fuels. As with most government policies, there were unintended consequences.

However, the government has a partner in the creation of the current agriculture and food system — you. Consumers spend more on agricultural commodities than any government does. The collective decisions of consumers have shaped the food supply into what it is today, with all its warts and issues. Collectively, society tends to choose cheaper, visually perfect, and convenient products. Hence we have a food system that produces cheap, highly processed foods and discards a lot of perfectly edible food that is misshapen or blemished. I understand the choices from the average consumer's perspective. Why would one pay more to buy a beef tenderloin from a local butcher when the big-box grocery store sells it for less? Most don't care that the lower price may be just an illusion from the perspective of their community as whole. The big-box store is paying their neighbours a part-time wage on part-time hours while hoovering money out of the local economy and sending it elsewhere, often across an ocean, to produce goods somewhere with lax labour and environmental laws. The best choice for a consumer and the best choice for society and for the future are not always the same.

This is where we come to the disconnect between the values of some eaters and what they see happening in agriculture. I say "some eaters" because the food movement is only a fraction of all eaters. The eaters involved in the food movement (a term I'm using loosely to roll up all the heterogeneous groups involved in the various dimensions of food activism) can see many problems with agriculture but are frustrated that "the system" stymies most attempts at reform. These activist eaters see a large disconnect between their values and the values reflected by agriculture.

Farmers, on the other hand, respond to what the marketplace asks them to produce. Record acres of corn have been planted in North and South America recently because the market has been paying record prices for corn. Farmers, first and foremost, are businesspeople trying to run profitable businesses to provide for their families and the families of their employees. The majority of us understand the intricate connection between what we do, the environment, and people's health. But if the market doesn't reward producing healthier food or food less damaging to the environment (or your definition of less damaging to the environment), then it is unlikely that farmers, as a whole, are going to produce healthier, less damaging food. People often ask why prices for organic are higher. The simple answer is: That's how much of a premium it takes to pull farmers into organic and produce food within the organic system in sufficient quantities to meet current demand.

Sometimes farmers' and society's interests are aligned. Acres being no-tilled (seeds being planted without tilling the soil) have generally been increasing across North America for the past 30 years. No-till reduces erosion, conserves soil, and reduces pollution of waterways compared to conventional tillage. From a farmer's perspective, those are all benefits, but the primary motivation has been the reduction in costs represented by no-till. A farm operation can run more land with less labour and less investment in equipment using no-till methods than it can using conventional methods. It just happens that society's interest in protecting the environment is aligned with the farmer's financial interest in this case.

However, a farmer's perception of his economic interests and the reality isn't always coincident. I endured ridicule (most of it behind my back) when I switched to organic methods. And I had some spectacular, very visible failures. I also had a number of fields that appeared to be failures from a conventional perspective that actually generated more margins per acre than any comparable conventional crop. The weed pressure had not diminished yields nearly as much as was assumed by other farmers doing the "80 kilometres an hour crop tour" from the seats of their pickups. When I moved two counties over, I often got, "Oh yeah, you're that organic guy who just moved into the Van Genechtin place."[11] I distinctly remember one gentleman asking me three years after we moved in whether the field by the county road was organic. He was surprised when I said yes. He was surprised because he thought it looked like the best field of grain in the township. I'm sure my answer was duly reported at the coffee shop the next morning.

---

11. This was despite the fact that the Van Genechtins hadn't owned the farm for over 20 years and two other families had owned it in the intervening years before we purchased it.

So how have I come to define sustainable? I've turned to the work of Allan Savory, creator of the Holistic Management (HM) system of managing agriculture. In HM, farms need to focus on four processes that all ecosystems have in common:

1. They are completely powered by solar energy.
2. They work with the water provided in the natural water cycle.
3. They recycle almost 100% of the mineral nutrients used within the system.
4. They are dominated by diverse populations of animals, plants, and microorganisms (primarily perennials).

The success of a sustainable farm hinges on its ability to manage and maximize those four ecosystem processes while minimizing the external inputs to the farm. That is the standard against which we need to measure any agricultural process or technology. If self-organizing ecosystems can create the vast forests, prairies, and marine ecosystems, we should be able to do better than the current destruction caused by annual crop agriculture.

Your decisions as eaters will have the greatest impact on how agriculture changes in the next decades. You can choose a sustainable future. But choosing sustainable agriculture involves more than choosing a particular brand or label at your local supermarket. It involves gaining an understanding of the myriad dimensions of sustainability in agriculture.

CHAPTER 2

THE TWO SOLITUDES

If we cannot envision the world we would like to live in, we cannot work towards its creation.

— *Chellis Glendinning*

A couple of years ago we had a lamb orphaned just before we went on vacation. Before being asked, a neighbour volunteered to bottle-feed the lamb while we were away. When we returned, we expected to pick up the lamb the next day. But when we called to arrange a time, we couldn't make the logistics work. After about a week, we finally just went over.

What we found was a family so enamoured with having a lamb around that they didn't want to give her up. She now had a name, Cauliflower, because the children thought that her tight curly fleece looked exactly like a head of cauliflower. More disturbing was that Cauliflower had imprinted on their dog. They chased each other around, chased cats together, and shared the dog's bed. Cauliflower had decided she *was* a dog.

When we brought her home, she chose our dogs over the sheep. Lilly was happy to have someone else to play with, but I'm not sure exactly what Kasne thought. Border collies like order in their world. I'm sure Kasne's thoughts ranged from "That lamb is nuttier 'n squirrel crap" to "What does this lamb think it is, sleeping with the dogs!" But he tolerated Cauliflower.

Meanwhile, visitors to our farm were delighted to watch the greeting party — a border collie, a guardian dog, and a lamb — come bounding down the lane. Excited lambs don't run. They bound. Bounding is like running but all four feet simultaneously touch the ground and then leap forward, except that young lambs sometimes aren't coordinated and jump straight up instead of bounding ahead.

At night, of course, our dogs do their security job, and they're very good at it. Even the cats don't get to cross from one building to another without the dogs reminding them to move quickly. But dogs have paws and lambs have hooves — the sounds they make crossing the deck below our bedroom window are very distinct. The first night I heard Cauliflower running behind Kasne to protect the farm, I almost split a gut laughing.

It was a breezy summer night. You could hear the yips and yelps of the coyotes and their pups out looking for dinner. Our dogs had dashed off to encourage the coyotes to try the buffet somewhere else, and Cauliflower was right behind them. I'm not sure what Cauliflower would have done when the game of tag she thought she was playing with her friends turned into a standoff with coyotes that saw dinner every time they looked at her. Can you imagine how confused a coyote would have been watching two dogs and a lamb running straight at it? "Are the dogs bringing a peace offering?" "Is it a trap?" "I'm so hungry I must be hallucinating — that third one looks just like a lamb!"

Cauliflower is grown up and part of our flock now. But I'm still not convinced she believes she belongs with the rest of the ewes, despite the fact that she has been penned and pastured with them for more than a year.

That's the problem with True Believers. Once they have an idea locked in their head, it is very difficult to dislodge it. We have a lot of True Believers in agriculture. There are True Believers in conventional agriculture. There are True Believers in organic agriculture. There are True Believers among animal welfare activists. There are True Believers in the anti-GMO crowd. There are True Believers in the pro-GMO crowd. True Believers are easy to spot — they generally prefer to loudly argue their position rather than engage in a meaningful conversation with people with opposing points of view. On many issues in agriculture, you have two solitudes that firmly believe they are right and continue to gather evidence that proves their position rather than taking the time to understand the opposing point of view.

I grew up a True Believer in conventional agriculture. When I bought my parents' farrow-to-finish confinement hog operation in 1995, we were "pork producers" raising pigs from birth to market weight. Our pigs only saw sunshine as they were being loaded onto trucks heading for the meat packers. You name the sin that modern agriculture is accused of, and I have not only committed it but also defended it.

The gradual change in my thinking started in 1998 when we planted our first certified organic crop. It was a purely profit-motivated move as we tried to avoid bankruptcy. A neighbour convinced me to give organic a try when I had the chance to pick up some land that was eligible to be certified immediately. We exited the hog business that year and focused on crop production. By 2005 we were 100% certified organic and, in an informal partnership with the neighbour, we ran over 2000 acres.

But I could see that we had some problems with fertility and soil structure. The organic standard says a nitrogen-fixing legume must be planted at least once every five years, but that isn't enough. The organic system really needs ruminant livestock (cows and sheep with four stomachs designed to digest hay, clover, and grasses) to ensure that the soil gets a rest under pasture or a hayfield for several years. So we decided to add cattle and sheep into our operation and reduced our land base by parting ways with our partner. Today we are focused on grazing sheep and cattle on perennial pastures.

As you will discover, I have questions about where conventional agriculture is leading us, but I'm not convinced that organic agriculture holds all the keys to the future either. I'm stuck in the portal between two parallel universes. Some of what I present in this book could get me tarred and feathered in organic circles. Other parts would draw heckles at some conventional agriculture events.

But first let me spell out a few definitions. I use *organic* to refer to the system of agriculture that is defined by the official regulations in various countries. The three primary rules of organic prohibit the use of synthetic pesticides, synthetic fertilizers, and antibiotics.

Quite a few people have heard me express the view that the worst mistake organic agriculture made was creating a codified set of rules that defined what was organic. The rules changed organic agriculture from a movement founded on the principle of reducing agriculture's impact on the environment to a set of rules to facilitate trade. Once the rules were established, producers' focus shifted to and inevitably came to rest on minimizing the cost of complying with the letter rather than the original intent of the rules. Codifying the rules also locked in one particular view of how agriculture's impact on the environment could be reduced. What's more, most of the viewpoints incorporated in the rules for organic originate from the early thinking of the movement, largely in the 1970s. A lot of new knowledge has accumulated since then. Unfortunately, the rules have essentially locked organic agriculture into using the technology deemed acceptable by the pioneers of the movement 40 years ago. Anything developed since is looked upon as potentially damaging.

Conventional agriculture is the term I apply to everything else. There is no official definition of conventional. It is agriculture free to use all the available technology to produce food and fibre. Some people refer to it as "industrial" agriculture, with "industrial" being used as a disparaging epithet.

Neither organic nor conventional necessarily implies or precludes the idyllic, overly romanticized setting that many urbanites equate with the term "organic." At our peak acreage, we ran 2000 certified organic acres. Looking around my yard, you would have seen a complement of farm equipment that matched or exceeded that on many conventional farms. I can take you to organic farms with tens of thousands of chickens and thousands of dairy cows. Is that a condemnation of organic? No. It reflects the reality that organic is subject to the same economic pressures to increase in size and lower costs as all other forms of agriculture. I can also take you to small, apparently idyllic farms that have problems with animal welfare and environmental damage.

When I was a salesman travelling the back roads of Ontario for one of the multinational agricultural chemical companies (now rebranded to a life sciences company), I made many calls on Mennonite and Amish farms that were governed by a host of rules about whether they could use technology like tractors or electricity. However, the rules rarely covered the use of pesticides. I remember one complaint call about one of our soybean herbicides. I arrived to find a horse-drawn pesticide sprayer with a ground-driven pump — an image that is jolting when compared to the idyllic setting that "Mennonite" or "Amish" evokes for most consumers.

Of all the arguments that define the two solitudes of organic and conventional, however, perhaps the most common one used to discredit organic as a long-term solution to feeding the masses is that organic agriculture yields less per acre and therefore would require more land to be cleared and used for agricultural production than continuing with high-input, high-yield conventional agriculture. All our views and analysis of most other environmental impacts will in some way be coloured by whether we are in fact trading yield for reducing a risk to the environment or human health when we choose between organic and conventional production methods, so let's pause to examine this overall argument.

In fact, a growing body of literature demonstrates that it is possible to achieve equivalent yields using organic methods. Truly long-term research was started in 1843 at Rothamstead, England, and continues to be funded by the estate of John Bennet Lawes, a successful entrepreneur and inventor of some of the processes used to make synthetic fertilizers who had a keen interest in understanding how they affected crop growth. At Rothamsted, some plots have received manure annually every year since 1852, some received manure from 1852 to 1871 and nothing since, and some have received no manure since the start.[12] In addition, four different rates of nitrogen (N) were applied to each of the manure treatments. In the no-manure and continuous manure treatments, there was virtually no response to the added nitrogen. From this we can conclude that when a lack of phosphorus (P) and potassium (K) are limiting plant growth, no amount of added N will increase yield. Similarly, a crop needs a maximum amount of N, and N can be adequately supplied by manure. Applying additional N did not increase yield.

The most interesting result was on the plots that received manure for 19 years at the start of the trial and then nothing for 100 years. In the yield data gathered in the 1970s, the plots with the highest rate of N came close to matching the continuously manured plots and there was a clear response to increasing rates of N. The data gathered in the 1980s showed a less pronounced response to N and the overall yields had declined relative to the continuously manured plots. There are two takeaways here. First, manure can supply all the nutrients required to maximize yield. Second, application of large amounts of manure can provide a long tail of available P and K. Not unexpectedly, nitrogen needs to be topped up much more frequently.

More recently, two long-term trials and two meta-analyses offer interesting findings. Since 1998, the Neely-Kinyon Long-Term Agroecological Research (LTAR) experiment at Iowa State University has run a fully scientific, randomized, long-term plot trial comparing conventional agriculture to several different organic rotations.[13] Averages from 13 years of this experiment show that yields of organic corn, soybean, and oats have been equivalent to or greater than conventional counterparts. A 12-year average for alfalfa and an eight-year average for winter wheat also show no significant differences between organic yields and the county averages. The organic plots were as productive as the conventional plots.

---

12. Wayne Martindale, "Rothamsted," 16.
13. Leopold Center for Sustainable Agriculture, "LTAR Experiment."

Unexpectedly, despite chemical nitrogen fertilizer being applied to the conventional plots and only nitrogen-fixing legumes and cover crops being used on the organic plots, total nitrogen (a key building block in protein) in the soil increased by 33% in the organic system — a finding that refutes the perception that organic agriculture perpetually suffers from lower nitrogen availability. Researchers also measured higher concentrations of other critical nutrients — including carbon, potassium, phosphorous, magnesium, and calcium — in the organic soils.

An 11-year trial at the University of California, Davis, comparing organic and conventional systems for tomato and corn production also concluded that soil quality, measured as total and protected organic carbon (carbon locked in stable or slowly decaying plant and animal material such as humus), can be improved using organic systems.[14] As well, tomato yields and the quality in the organic system were both equal to or better than in the conventional system, though corn yields were lower because planting was delayed to effectively kill weeds in the organic system.[15]

In 2004, Canadian researchers released a meta-analysis summarizing the state of global research on organic systems.[16] They reported on studies that concluded organic was equal to conventional in tomatoes, apples, and dairy, and within 5% in corn soybeans and wheat. Though all these trials demonstrate that yields in organic systems can match yields from conventional systems, the organic compost used in the trials was an external input, just as the fertilizers used in the conventional system were. Thus, apologists for conventional agriculture argue that if the acres used to produce feed for the animals to produce the manure and the acres used to grow the nitrogen-fixing legumes in the organic treatments were included, the average yield would be substantially less.

Using evidence from 293 articles that reported comparisons between organic and conventional yields, a team of researchers from the University of Michigan looked at this very question.[17] From these articles, and a model that included allowances for legumes to fix the required nitrogen, they projected world food production using both conservative and optimistic yield scenarios. They concluded that, even

14. A. Y. Y. Kong et al., "Carbon input, aggregation, and soil organic carbon stabilization."
15. "Crop Yields," Russell Ranch Sustainable Agriculture Facility, Unveristy of California, access July 8, 2013, http://ltras.ucdavis.edu/res/crop-yields.
14. Rod MacRae et al., "Adoption of organic food."
17. Catherine Badgley, "Organic agriculture."

under the conservative scenario, organic methods have the potential to match conventional agriculture's calorie production per person.

This study has attracted criticism from conventional agriculture scientists for a number of reasons, including using studies from the non-peer-reviewed "grey" literature, categorizing some treatments as organic despite not conforming to the strict definition of certified organic, and counting yield results from the same trials multiple times because they were included in multiple papers. These are all valid but nit-picky criticisms. Given the number of independent trials considered in the Canadian meta-analysis, I doubt that conclusions reached in the University of Michigan study are wildly inaccurate — the yield ratios used are very similar to the Canadian findings. Should the estimates be considered precise? No. However, they demonstrate that the yield potential of organic agriculture is within the same range as conventional agriculture.

Finally, *Global Development of Organic Agriculture: Challenges and Prospects* by Niels Halberg et al concluded that organic agriculture is a valid system for improving the food security of the developing world, a direct repudiation of the view that food insecurity in the developing world is a result of not using advanced agricultural methods and technology. The use of locally sourced fertility sources within an organic system resulted in greater food security than a conventional system, which required the export of food to generate income to purchase pesticides and fertilizers.

All the research that I have referred to is focused solely on the straight-up comparison of organic to conventional yields. The research did not include any projections of what the production environment might look like in 50 or 100 years if we continue with the fossil inputs of conventional agriculture.

And there's something else that you need to consider as you review this and other research. The studies are comparing a conventional system that has been developed, optimized, and advanced by the collective knowledge of almost every agricultural researcher on the planet for the past 50 years. The organic systems are replications of systems that farmers have developed by relying on the knowledge of previous generations but without the benefit of the agricultural establishment. Would the results improve if similar efforts were expended on organic research? Could the combination of traditional systems with the modern understanding of the details of biological systems produce even better results?

For example, crop-breeding work is focused on producing crops that have higher yields in conventional production systems. Plant breeders are selecting crops that perform well in high-input systems. But it is possible that we are losing traits that would help crops perform well in low-input systems. So far, most research and my personal experience indicate that varieties that come out on top in research trials in high-input systems also come out on top in low-input systems. However, now that crop-breeding research is focused primarily on genetically modified crops, the number of new varieties allowed for organic production is declining.

Another angle to examine when comparing the yields is the scale of the systems being compared. In agriculture school, when we looked at the concept of economies of scale, the analysis started at 100 acres and went up from there. Production efficiencies are gained until you get to a farm size that keeps the biggest machines working. Currently in North America, most economies of scale are captured by the time you get to 5000 acres. But those are production efficiencies in terms of cost of production, not yield per acre. It is easy to see that the small CSA gardener produces a much higher yield of vegetables per acre than a monoculture grower of vegetables. This is borne out in a study recently completed by the UN's Food and Agriculture Organization (FAO) examining production methods in the poorest nations. What they found was that when you looked at production in smallholdings, the yields went up as the holdings got smaller — the inverse of what was inferred by extrapolating backwards from 100 acres on the North American curve. They also found that the smallholdings used primarily organic methods, while none of the studies I referred to above made any allowance for an intensification of production through increases in smallholder production.

Time is the final dimension to be considered. Conventional agriculture is feasible today because we have abundant, relatively cheap fossil fuels and mineral resources. Those resources, however, are finite. At some point, agriculture will be pushed to reduce fossil fuel inputs, fossil mineral inputs, and fossil water inputs because their increasing scarcity will increase their prices, and the increasing levels of pollution will force increased regulation. Methods used primarily in organic agriculture today will be incorporated into conventional agriculture in the future, along with technologies not even conceived of yet. Can I prejudge whether these new technologies will improve the intersection between agriculture, ecosystem, and human health? No. Some will. And some won't.

In the following chapters, I will walk you through a number of issues where the sustainability of agricultural practices and their impact on the environment are cause for concern. However, I don't think we are forced to choose between feeding people and reducing our impact on the environment. In fact, reducing our impact on the environment is key to sustaining food production long term. That said, "reducing our impact on the environment" and "organic" are not synonymous. One of the bigger challenges we have to overcome is engaging the two solitudes of organic and conventional agriculture in a discussion for the long-term benefit of agriculture and eaters.

CHAPTER 3

THE DIRT ON DIRT

The nation that destroys its soil destroys itself.

— *Franklin D. Roosevelt*

Have you ever stood in an old-growth forest and marvelled at the breadth of the trunks and the height of the trees? At the root mass that anchors each tree? Most plants have as much biomass under the soil as above it — and giant trees are no exception.

This is what an ecosystem can accomplish when people don't interfere. No one disturbed the soil to plant the original seed. No one watered the seedling. No one used fossil fuels to create nitrogen fertilizer and pesticides to feed and protect the tree. No one mined the phosphorus that is in the DNA and powers the energy cycle of every cell in the tree. No one sprinkled potash around the base of the tree to help it pump water up the great trunk. Natural ecosystem processes accomplished all that, the majority of them occurring out of sight in the soil.

Soil is not a homogeneous substance but a complex ecosystem that exists in between the fine rock particles (in order of size from largest to smallest these rock particles are called gravel, sand, loam, and clay), decaying organic matter, air, and water. "Dirt" is a thriving metropolis of microscopic and not-so-microscopic organisms. And each of these various organisms has a vital role to play in the soil's ecosystem. Some of them decompose organic matter from dead plants — both above- and below-ground biomass. There are predators and prey. There are scavengers. There are symbiotic relationships. There are parasitic relationships. The soil is a microcosm of what we observe in human-scale ecosystems. When you scoop up a handful of rich soil, you have more life forms in your hands than there are human beings on this planet.

That is how full of life soil is. The "earthy" smell of the soil, for example, comes from a genus of fungi called actinomycetes.

When you look at an unspoiled ecosystem, there is an energy cycle, a water cycle, a mineral cycle, and a life cycle. Soil is the intersection of all of these cycles, and the soil ecosystem is the largest recycling operation on the planet. Minerals are taken up from the soil through the roots of plants, eaten by animals, and deposited back on the soil in animal excrement or dead plants. The excrement and dead-plant material provide energy for the "decay" organisms that cycle the minerals back into plant food, as well as extract additional minerals from the soil and the air. The soil digests the biomass and uses the stored energy to digest minerals in the soil and supply them to the next generation of plants. Soil rich in organic matter also stores significant amounts of water to nourish plants between rains. While the energy to grow a tree comes from the sun, just about everything else, including much of the water and all the minerals, comes from the soil. Without a properly functioning soil ecosystem, all the biomass created by plants just builds up.

Soil supplies minerals for life, but it is also a creation of life. Rich, dark garden soil — what we refer to as "topsoil" — didn't exist before life on this planet. Topsoil was formed as plants grew, died, and decayed over long periods. If you live in a subdivision built in the past 50 years, you will have soil — often referred to as "hard clay" by urbanites — in which it is almost impossible to grow a vegetable garden without adding a lot of compost and peat. This soil is actually subsoil — the soil below the layer where organic matter accumulated and mixed. If you have ever tried to garden in subsoil, you will understand why it is so important to maintain topsoil to maximize food production. But while you can "make" enough soil to grow a small garden on your urban lot, when we scale up to the size of a farm, building soil is a very slow process that takes decades and centuries to measure appreciable increases. We can change some of the properties of our soils over the course of five to ten years, but the accumulation or creation of new soil is measured in centuries.

Nature has many ways of holding soil in place. Do you know what the largest living organism on earth by both mass and volume is? It's not an elephant. It's not a whale. Its botanical name is Armillaria solidipes and its fruiting bodies are commonly known as honey mushrooms. A single fungus of this species covers at least 2200 acres in Malheur National Forest in Oregon and it is estimated to be over 2400 years old. This is a fungus that has been allowed to grow without disturbance. It is hypothesized that it has been able to survive and spread because the dry environment is hostile to other fungi establishing themselves and competing for nutrients, but the fungus is responsible for the death of large patches of pine trees. There is another member of the same species in Washington State that covers a paltry 1500 acres. The strands of these fungi and substances they excrete are an important "glue" that holds soil together.

Unfortunately, soil can be lost, or "eroded," much faster than it accumulates. Paradoxically, all of our soils were created by the forces of erosion grinding on the rocks that constituted Earth's surface to create the mineral fraction of the soil that now provides life to the majority of vegetation. These same forces, however, work to wash and blow soils away when they are left exposed, making erosion an important yet imperceptible threat to our soils.

In the Great Lakes Basin, average soils have the potential to lose between 6 and 11 tonnes per hectare (2.7 to 4.9 tons per acre) per year. The average 100-acre farm has the potential to lose eight tractor-trailer loads each year. Eight tractor-trailer loads every year. That amount is at once a very large number and a very small number — it depends on your perspective and time horizon. Eight tractor-trailer loads look like quite a pile of soil when they're dumped on a suburban lot. It would be over 1.3 metres (4.3 feet) deep. A raised bungalow would no longer be raised. An average male would be buried chest high. When you spread it across 100 acres, however, it's about 1 millimetre (0.04 inches) thick, imperceptible to a human eye, in the space of a year. In the space of a century, it is 10 centimetres (4 inches). In the space of a millennium, it is over a metre.

The math in the preceding paragraph applies when soils are eroded and deposited in waterways and lost to production, but not all eroded soils leave agricultural production. Some just move from the top of a knoll into the flat below. Some blow from one farm to the next. Even the soils swept away by rivers are not necessarily lost to production forever. Some of our richest soils globally were deposited at the mouths of rivers to form river deltas. The flooding of the Nile was part of the desalination process and contributed to the deposition of fresh nutrients in the soils of the Nile Valley. However, it is irrefutable that we are losing soil to erosion. What people who take a less sensationalistic view of erosion are arguing about is the timeline. If you look at the earth from a 10,000-year timeline, humans have inhabited North America for at least 10,000 years.[18] For the first 9700 years, soil and organic matter were accumulating. For the last 300 years, they have been diminishing.

Everyone who farms knows it's happening. We all know that drainage ditches silt up and need to be cleaned out at least once a decade. We've all seen little cuts of rill erosion on the sides of hills after a heavy spring thunderstorm. Scientists can measure how many tonnes of soil are blowing off the major agricultural regions of China and ending up in the Pacific Ocean because the dust cloud is measureable in satellite images. And the signs are already visible when you drive around the countryside. If you've ever looked across a tilled field and noticed that the knolls (mini hills) were a lighter colour, that's a sign of erosion. The light soil is subsoil. All the topsoil has eroded off the knoll, some by water, some by wind, and some by tillage. Yes, the tools we use to till the soil gradually move soil downhill. Tillage tools lift and throw the soil to shatter it, allowing gravity to pull the soil slightly downhill. If you've ever raked out gravel to level it off, it's the same process.

Left undisturbed, soil forms very stable aggregates that are held together with biological glue. This "glue" is a combination of a number of factors: the filaments from actinomycetes, the polysaccharides they secrete, the slime that slugs and related organisms secrete, and chemical reactions between neighbouring particles of soil. The longer the soil is left undisturbed, the more stable the aggregates become. If I plough soil that hasn't been disturbed for many years, it is very different from soil that has been disturbed annually. Standard agricultural practices disrupt and destroy the soil ecosystem that is critical in the prevention of erosion. Both conventional and organic agriculture use tillage to churn up the top 10 to 15 centimetres (4 to 6 inches) of soil at least twice a year. Can you imagine the impact this has on the thin filaments of actinomycetes that are spread throughout the soil? What happens to the organisms that live on the surface and depend on the surface biomass when they are suddenly buried?

Conventional agriculture cares little for the soil beyond the level of nutrients it contains — and even then the focus is really on only three elements: nitrogen, phosphorus, and potassium. A common practice

---

18. Archaeologists disagree about whether the archaeological record indicates more than 10,000 or closer to 16,000 years ago. Most agree that the so-called Clovis people were widely distributed in North America at least 11,500 years ago. Another line of evidence suggests that a civilization predated the post-Ice Age Clovis by at least 1000 years.

when growing high-value crops (vegetables, strawberries, tomatoes) is to fumigate the soil before planting — essentially killing all the life in the soil to ensure that a few fungi that attack plant roots are killed. At least with the fumigant we know that we are killing everything. We do very little research on the impact of pesticides on the life in the soil, partially because we know very little about the complex ecosystem that exists there. What is the impact of all the GMO corn stalks full of the Bt insecticide?[19] If you're a member of the Lepidoptera order of insects (moths and butterflies), potentially quite a lot.

Most of the organisms in the soil ecosystem can only operate in the dark, moist environment of the soil. They can't run up a stalk of grass and take a bite from the top. They have to work on what is in the soil. This brings me to an important member of the soil ecosystem that is easily seen with our eyes — the large herbivore: cattle, deer, moose, water buffalo, elephant, and rhinoceros. In a straight no-till system (seeds are planted with no soil disturbance and weeds are managed with herbicides) without a large herbivore, we see nutrient stratification. The nutrients tend to build up in the surface layer as the plant matter slowly decomposes. If the field is in an arid environment, the plant matter doesn't decompose — it oxidizes without the soil life benefitting. If you introduce a large herbivore into the system, it does several things. First, the animal eats the plants, chews them, and ferments them in its rumen, doing a lot of the initial complex digestion, and then deposits the plant matter in a moist patty on the soil surface. The large herbivore also tramples the plant matter into the top layer of soil. The recycling of nutrients is a lot more efficient with a large herbivore in the system. They transport nutrients back up hills and move it from the margins of forests to the centre of meadows. If you look in almost every terrestrial ecosystem, you will find a large herbivore. And the drier or more "brittle" the region is, the more important the large herbivore is to nutrient recycling. Its rumen may be the only moist place where the plant matter can be broken down by fermentation.

Undisturbed, this is the only ecosystem on the planet that builds soil. This ecosystem created the depths of soils that we have. Plants grew using energy from the sun and then were either eaten by herbivores or they died

---

19. "Bt" corn is corn that has had a gene from the genome of Bacillus thuringiensis bacteria inserted in the corn's DNA to cause the corn to produce a toxin that is toxic to specific genera within the insect phylum, particularly within the Lepidoptera order.

and were recycled into the soil to provide nutrients for the next plant. Over millennia, organic matter built up in the topsoil, and these are the fertile soils that humans plant crops in. As a society, we are mining a lot of "fossil" resources: fossil fuels, fossil water (depleting aquifers and glaciers), and fossil minerals. We're also mining "fossil" organic matter. Undisturbed virgin soils that had levels of organic matter of 5% to 10% have been reduced to fewer than 2% because of tillage and continuous cropping.

Organic matter is an excellent buffer in adverse conditions. It holds up to 90% of its weight in water — water that is available to plants during dry spells. Organic matter releases nitrogen in a plant-available form in a pattern that mimics most plants' needs for nitrogen (slowly in the cool spring and then accelerating rapidly as soils warm up). And organic matter helps stabilize soil aggregates and mitigate erosion. But organic matter is very difficult to build back up after it has been lost, especially in a system that involves continuous cropping with annual crops.

One of the key principles in organic production is to always be working to build your soil. The healthier your soil is, the healthier your plants and your animals will be. There are three elements in the certified organic standards designed to ensure that organic farmers are building their soils. The first is the mandatory inclusion of soil-building legumes a minimum of one year in five in your crop rotation. The second is a prohibition on summer fallowing as a management practice. Summer fallowing involves leaving a field fallow (bare) for an entire season and tilling it every time you start to see weeds growing, usually about every two weeks. Summer fallowing kills perennial weeds, but it also destroys soil structure and burns or oxidizes organic matter. The third way organic farming conserves organic matter compared to conventional agriculture is the lack of high application rates of nitrogen fertilizer. The process of breaking down organic matter requires nitrogen, but when excess nitrogen is available, it will accelerate the decomposition of organic matter.

Organic agriculture uses soil disturbance to kill weeds, however, and disturbance destroys soils. There is a long-running argument over whether the tillage damage done by organic agriculture outweighs the benefits of the soil-building practices. Conventional agriculture also points to no-till as a system that leaves soil undisturbed while growing crops. In 2007, J.R. Teasdale and his colleagues at the Agricultural Research Service, the in-house research arm of the US Department of Agriculture, reported on a nine-year comparison of conventional no-till, traditional certified organic methods, and two intermediate systems. They concluded, "These results suggest that organic can provide greater long-term soil benefits than conventional no-till, despite the use of tillage in organic."[20] The research also found a higher concentration of carbon and nitrogen at all depths up to 30 centimetres (1 foot) in the organic treatment compared to all others.

These trials were conducted on erodible droughty soils to provide the greatest challenge for the organic system to build soil. And while this study doesn't prove that organic methods are better soil builders, it does show that well-managed organic methods can match and surpass conventional no-till in both long-term carbon sequestration and soil nitrogen content. However, both systems reduce but do not eliminate erosion. And just because organic has the potential to reduce erosion doesn't mean that it systematically does. I confess that I caused more erosion in my first years as an organic grower than I did in my last years as a conventional grower.

I searched and struggled for a decade with the problem of soil erosion in organic agriculture. Most of the erosion occurred in soybean fields. Soybeans were an important part of the economics of our farm but a weak competitor against weeds. Organic soybeans require several passes of tillage before planting to kill as many weeds as possible. After planting, we do an additional two to five passes to kill the weeds that germinate. By the time soybeans are tall enough to shade any weeds, the top of the soil has been turned to dust.

Trying to deal with this problem, I was one of the first certified organic farmers in Canada to experiment with organic no-till production. The system involved planting a crop of rye in the fall, waiting for it to head out in the spring, rolling it flat into a mulch mat, and then planting the soybeans. The combination of the rye straw mulch and the allelopathic chemicals released by the rye straw as it breaks down will suppress the weeds and the soybeans will grow well without any tillage.[21] The Rodale Institute in Pennsylvania pioneered the technique and was successfully using it with both corn and soybeans.

We encountered a number of challenges. First, to ensure that the rye crop is thick and robust enough to shade the ground in the spring and prevent any weeds from germinating, the rye crop needs to be fertilized. We had a limited livestock base and our manure was better utilized right before the spelt crop to ensure the highest-quality spelt. Second, we have a

---

20. J.R. Teasdale et al., "Cover Crops."
21. Allelopathic chemicals are chemicals released by one organism that impact the growth of another. In this case, rye straw releases a chemical that inhibits the germination of other plants.

shorter growing season than in Pennsylvania and waiting for the rye to head can leave planting the soybeans until well after the optimum date. Finally, a point that was unique to our operation, "volunteer" rye in subsequent crops is unavoidable and our operation has spelt as an important crop and a very low tolerance for contamination of rye in the spelt.

In the end, we moved on from organic no-till. To reduce erosion and soil depletion, we removed soybeans from our rotation and focused on spelt, hay, and mixed grains (the latter a mix of oats, peas, and barley to supply our own chicken and pig feed). With this rotation, we were able to keep 100% of our ground covered through the winter and there were only short intervals when planting spring grains and spelt where the soil was bare and open to erosion. However, removing soybeans leaves a small dent in the diet of our chickens and pigs. We can mostly use peas to replace the protein in the mixed grain, but soybeans supply a couple of amino acids, lysine and methionine, that animals cannot synthesize themselves and are not available in sufficient quantities in many crops. In an interesting twist, amaranth is a good source of lysine, and quinoa is a good source of lysine and methionine. And two of the most common weeds we fight are pigweed —a member of the amaranth family — and lamb's quarters — a member of the Chenopodium family, as is quinoa. I'm hoping that all those "weed" seeds that "contaminate" my feed grains are actually providing a reliable source of the missing amino acids.

We've now moved to primarily livestock agriculture with a heavy emphasis on ruminants that can be pastured. We have our entire land base in permanent pasture and are using rotational grazing methods described later in the book to actually enhance and create soil.

But now the question becomes, since we can't live on meat alone (as much as my three teenage sons would protest to the contrary), how do we round out our diet? Most of the fruits we eat are perennials. On our farm, we are gradually establishing plantings of apples, pears, plums, peaches, blackberries, rhubarb, raspberries, and currants. If you look at most sources on permaculture, you will find that the diet's energy source comes from nuts, so we are also establishing walnuts, hazelnuts, pecans (hopefully), chestnuts, and low-tannin oak trees for edible acorns. All these trees are being planted in our fencerows, and in some cases where fencerows used to be, to help hold soil on slopes, provide windbreaks, and provide shade for our livestock.

Still, I don't think we have to go to 100% permaculture systems to prevent the long-term destruction of soils. A system in Argentina shows a lot of promise. In an eight-year rotation, five years are intensively managed pasture and three years are annual crop production. When you give the soil five years of undisturbed building, the soil aggregates are much more stable and the soil is less prone to erosion when it is tilled for the annual crop production. Additionally, the forces of water erosion are greatly slowed if the water flows off tilled soil onto pasture rather than into ditches and waterways because most of the soil that was moving in the water settles out in the pasture. This type of rotation makes logical sense to me, but it is so uncommon here in North America that I can't find much research that tests my hypothesis. And unfortunately, because of rising grain prices, the Argentineans are switching their rotations for something that looks a lot closer to North American annual crop rotations.

Another approach to perennial agriculture may come from plant breeders and gene splicers, who are working hard on developing a perennial grain crop that rivals annual crops in yield. Perennial grasses are not nearly as prolific as annual grasses at producing seed because a larger portion of their energy is devoted to growing and storing energy in the root system to survive through the winter and regrow the following spring. Corn, wheat, barley, oats, and spelt are all annual grass crops. The researchers are getting close to a new perennial grain. Most of the work to date has been done using traditional breeding techniques, and one researcher has bred a wheat/wheat grass cross that is a perennial and yields 60% to 70% as much as modern wheat varieties. I have a small plot of perennial rye grass selected for seed production. My plan is to continue to select for seed production and hardiness for the next decade and see whether I can come up with something that has a good enough grain yield to plant on a larger scale.

All these techniques are focused on reducing the loss of our existing soil. But in some places we are already at the point of needing to rebuild. The reading I've done suggests that it takes between 200 and 700 years to build an inch of soil. If that range is correct, it could take 2000 years to replace the soil I can lose to erosion in the length of my farming career. That's right: If one of my ancestors living during the time of Christ caused erosion at a similar rate to what modern agriculture does, and if that land was left fallow from when he died until now, nature would just now be finishing rebuilding the soil that was lost. That's a little sobering. Even if the number is out by a factor of ten and it takes 50 years to rebuild an inch of topsoil, we're talking two centuries of undisturbed soil building by natural processes to rebuild the soil lost during the career of one farmer.

It's on the erosion dimension of the sustainability problem that I attack organic and conventional equally and see the greatest threat to future generations' ability to feed themselves. There is no quick technological fix for soil washed into the Gulf of Mexico. The longer we continue to allow soil to wash away, the bigger the dip in food production is going to be in the future. The longer we burn organic matter with high nitrogen inputs, the less food we will have in the future. We have our collective foot firmly pressing the accelerator of food production to the floor. We're screaming down the highway redlining. We feel like we're powering ourselves into the future, but the longer we have it matted, the more damage is being done to the engine. At some point we will throw a rod and come to an abrupt, sickening halt. And up to the second before the rod is thrown, we will be oblivious to the impending doom. I know, because I threw a rod in a swather engine. It came right out the side of the block of an old indestructible six-cylinder Ford. Turns out all the oil had leaked out and the oil pressure sensor wasn't working. We know better than to run down the highway with the engine screaming, but we're doing it anyway.

Parts of North America will run out of topsoil before the planet runs out of oil. If we continue to lose soil at current rates, my grandchildren will witness the last of our topsoil washed or blown away. My grandchildren. That's not some esoteric geological time horizon. That's real. That's people who will be at my funeral. That's people who will see what's happening and ask me why I made the choices I did. Questions like, "Why did you choose not to use pesticides and fertilizer but allow the soil to wash away?" Tough questions. Hard questions.

Many of you will see pesticides as the greatest threat posed to the environment by agriculture. Or maybe you line up against phosphorus loading of our lakes and waterways. From where I'm sitting, erosion is the one problem in today's agriculture that, if left unaddressed, will most assuredly lead to a diminished ability to feed ourselves in the future.

Whether the 10 centimetres (4 inches) of soil loss in the next century is meaningful depends on your endowment of topsoil. It's not much if you're sitting on several feet of topsoil in a river flat somewhere, but most of us in Southern Ontario farm on 15 centimetres (6 inches) or less of topsoil.

But you don't consume soil. You don't know if the choices you're making at the supermarket help lose or save soil. You can minimize food miles and still end up with barren landscapes with insufficient topsoil to grow healthy crops. Of course, you can always claim ignorance — you just shopped at the stores. You can tell yourself the government should have done something about industrial agriculture destroying the environment.

But wait. I'm an ethical, certified organic farmer, and it happened to me too. It isn't just about the industrial farmers. It's never about just one dimension. You can choose a certified organic meal every day from now until the day you die and you will have missed this one. A utopian future where all the available land is dedicated to annual crop production using organic methods is not in fact a utopia. Organic agriculture replaces herbicides with soil disturbance and soil disturbance leads to erosion. Eroded soil deposited in the lakes and oceans grows no crops.

There is no magic label that says "erosion free," so how can you as a consumer affect soil erosion? The simplest shifts are in the content of your diet. Eat more perennials, such as fruits and nuts (peanuts don't count). Eat grass-fed meats. Pasturelands are planted to a diverse mix of perennial crops. Even if the pastures are broken up every five to ten years to plant grain before being returned to pasture, that is far better for the soil than continuous annual cropping.

To get more sophisticated and make an assessment of whether the rest of your diet is causing erosion, you need to be able to ask the farmer growing it some questions. In my mind, this is the biggest benefit of locavorism — the ability to observe and question the farmer. There are three questions you need answered to give you a reasonable understanding of the erosion potential on the farm growing your food:

1. What crop rotation does the farm use?

   What you're looking for are periods, measured in years, where the soil is not tilled or disturbed. It will primarily be some form of hay or pasture that is planted during this period. No-till practices are acceptable too, but understand that herbicides will be involved.

2. What soil-building methods does the farm use?

   A good answer will involve frequent use of cover crops. Fields should not be left bare any longer than is absolutely necessary. Of special consideration is the use of cover crops over the winter in northern climes and during the hot, dry, non-productive season farther south. The best protection for soil is a living cover crop. Blowing snow and spring runoff are prime times for erosion.

3. How are the tilled parts of the farm managed to reduce erosion?

The best answer is that no tillage is done, either through no-till planting or the use of perennial crops. Most farms growing crops other than corn, soybeans, and wheat will do some tillage. Meanwhile, water erosion is primarily powered by gravity. The longer the undisturbed path is down a hill, the more damage water will do. Steep slopes should not have annual crops. Moderate slopes should be managed using a technique called "contour cropping" where the crops are planted across the slope and row crops (corn, beans) are alternated with strips of hay or pasture to catch the water, slow it down, and settle out soil. Wind erosion is best managed by slowing the wind down with windbreaks and cover crops.

If you have an opportunity to visit the farm growing your food, examine the fields. Can you see the perennial crops in the rotation? Do you see a lot of bare soil? Do you see places where water has cut across a field? Are there fencerows or is the farm one big field? You can choose to stop the erosion of our future — careful observation and common sense will tell you almost everything you need to know.

CHAPTER 4

FERTILE LIES

I would feel more optimistic about a bright future for man if he spent less time proving that he can outwit Nature and more time tasting her sweetness and respecting her seniority.

— *E. B. White*

My mother was a city girl through and through, but her father had an innate understanding of the fertility cycle. He would go into the street and gather the "road apples" after the milkman on his horse-drawn carriage had delivered the milk or any other horse had passed by and left their calling card. Living in the heart of Toronto, collecting animal manure to put in your garden wasn't exactly commonplace. There certainly wasn't anyone in the neighbourhood racing him for the latest deposit. Like dirt, manure wasn't something that people desired to come in contact with. Yet, that manure is an important part of recycling minerals. My grandfather used that horse manure to add fertility to his garden.

The wild ancestors of pigs were an important part of the fertility cycle in many ecosystems. They consumed living and dead plant material, small animals, carcasses, and essentially anything else that struck their fancy. In the process of seeking out their meals, they rooted through the litter covering the soil and mixed in the organic matter along with seeds that they missed consuming. They digested all of this material and returned it to the soil as a more plant-available source of fertility. Pigs were the original composters, if you will.

Using our farm pigs, this fall I took a "unique" approach to cleaning up Silvia's garden and ridding it of some weeds. Silvia's garden has been gradually overrun by six-foot-tall sunchokes — a member of the sunflower family, native to North America. Sunchokes don't produce large seed heads like most sunflowers. Instead, they store carbohydrates in tubers and generally reproduce from their tubers or even pieces of their roots. Once you have them in your garden, they are very difficult to eliminate completely without a herbicide. However, the tubers have a slightly sweet and almost nutty flavour. Once we had harvested what we could from the garden this past fall, we decided to turn some pigs in. We figured that once they discovered the sunchoke tubers, they would root up the entire garden looking for them. The pigs would do three jobs for me — work over the garden, eliminate the sunchokes, and apply some fertilizer.

We started with two pigs that were partially grown and had already spent a couple of months "working" with their littermates in a corner of a pasture that needed to be worked up and replanted. These two went through the garden and ate all the remaining cucumbers, tomatoes, peppers, and small squash, but they didn't seem interested in working for their supper. I even rooted-up sunchoke tubers for them in case they hadn't discovered the tubers themselves. But instead of eating the exposed tubers, they escaped the garden and started rooting up the lawn with great glee. (Yes, pigs demonstrate glee. They make a sound that can only be described as a combination between a bark and a snort while simultaneously running in circles.) Two pigs can roll up a sizeable chunk of lawn in a very short time. Their noses are perfect tools for peeling the sod off the top and then rooting through the dirt.

We decided to bring in reinforcements. The next four pigs had no trouble finding the sunchokes, methodically unearthing them from the entire garden. (The original two had no trouble helping themselves to what the newcomers rooted up.) By the time the pigs were ready to become bacon, Silvia's garden had been completely cleared of the year's leftover plants. A high percentage of the sunchoke parts were consumed, and the garden was ready to go into the winter. The six-foot-high jungle of sunchokes had been chewed up and mixed into the soil. I hadn't spent a dime on diesel, although I might have to buy some grass seed in the spring.

The process I used to rid Silvia's garden of weeds was essentially the natural soil-building process in an extremely concentrated and focused form. We made a time-limited, intensive use of our pigs. If we left the pigs there all year, that soil would eventually be destroyed. They would prevent any plants from growing to cover the soil and prevent erosion, and they would continually stir up the soil, causing organic matter to be lost. Their hooves (pig's feet are essentially four-inch stilettos supporting

200 pounds of bacon) would create a layer of compacted soil that would limit the ability of plant roots to grow further down into the soil. Pigs can be a useful tool, but use too many or leave them for too long and they will wreak havoc. That is the paradox of most tools in agriculture — they can be helpful, but they can also be destructive.

In agriculture, fertility is always applied to the fields in some form to replace the minerals that are shipped off the farm in the form of food. Synthetic fertilizers are one of the key differentiators between organic and conventional food production. There are three macronutrients that are applied as fertilizer: nitrogen (N), phosphorus (P), and potassium (K). Organic farmers are limited to natural source fertilizers, such as compost, naturally fixed nitrogen from leguminous crops, mined rock phosphate, and mined potash. Conventional agriculture, on the other hand, relies primarily on synthetically compounded nitrogen, phosphorus salts created by treating the mined phosphorus with acids, and potassium mined as potash. The remainder of this chapter discusses the different approaches to supplying each of the macronutrients and the sustainability of each.

NITROGEN

Nitrogen is a key building block in both plant and animal life. Amino acids are differentiated by the attachment of an amine group (NH_2). Amino acids are the key building blocks of protein. Without nitrogen, there is no protein. Nitrogen fertilizers are created by taking nitrogen gas (N_2) from the air and reacting it with natural gas (CH_4) under pressure and with heat in the presence of a number of catalysts to create ammonia and then urea and ammonium nitrate fertilizers. This process is generally referred to as the Haber-Bosch process and was discovered initially by Fritz Haber in Germany at the beginning of the twentieth century. Carl Bosch scaled up the process to create an industrial, continuous flow process capable of generating tonnes of ammonium nitrate, primarily for munitions in the First World War. At the conclusion of the First World War, the nitrogen factories turned their attention to producing fertilizers for agriculture.

Nature has its own paths for "fixing" gaseous nitrogen (N_2) — 80% of the air we breathe — into a form that can be used by plants. The incredible energy and heat released during a lightning strike fixes vast quantities of nitrogen that fall to the earth. There are also a number of

47

organisms living in the soil and water that are capable of fixing nitrogen. Primary among them are the bacteria within the *Rhizobium* genus. These bacteria form symbiotic relationships with legumes. The bacteria live in nodules on the roots of the legume plants. The bacteria fix nitrogen and supply it to the plant in exchange for a supply of sugars and minerals.

Conventional agriculture is dependent on fossil fuels for fixing nitrogen. The single largest consumption of fossil fuels in the production of corn is not the diesel fuel consumed by the tractors, combines, and other machines used in the production of corn. It's the consumption of natural gas in the production of synthetic nitrogen fertilizers. The price of nitrogen fertilizer follows the price of fossil fuels fairly closely.

Our friends in the biotechnology world are trying to genetically modify non-legume plants to form root nodules and symbiotic relationships with Rhizobium bacteria. A second line of research is working on inserting the genes for nitrogen fixation into the major nitrogen-consuming field crops. Both processes are complex and governed by a number of genes. Plus, the nitrogen-fixing process requires a significant amount of energy regardless of whether we are talking about the man-made or biological process. One of the reasons the Rhizobium bacteria are efficient nitrogen fixers is that they focus on fixing nitrogen while the legume plant supplies them with energy.

The good news is that regardless of whether research comes up with a genetically engineered solution or we diminish our natural gas reserves to the point that prices escalate, we already know how to manage the natural nitrogen cycle through legumes to provide for the nitrogen needs of subsequent crops. The techniques and knowledge being accumulated by organic farmers will be critical to supplying the nitrogen needs of crops in the future.

However, the continuous application of synthetic nitrogen fertilizers diminishes our future ability to grow food. First, synthetic nitrogen fertilizers "burn up" the organic matter in the soils. Nitrogen is required as part of the microbial process that breaks down long-chain hydrocarbons in the soil. This is a sword that cuts two ways. When you incorporate a substantial amount of carbon biomass into the soil, the decomposition process sucks up available nitrogen. On the other hand, if you have surplus nitrogen, it will work to degrade the organic matter faster. All of the organic sources of nitrogen come balanced with organic matter, whether it is a legume for ploughing under or animal manure or compost. Excess nitrogen burning up organic matter not only reduces the productive capacity of the land, it releases carbon into the atmosphere that had previously been sequestered in our soils. This in turn increases the atmospheric concentration of greenhouse gases and contributes to climate change.

Synthetic nitrogen fertilizer contributes to all the ills associated with the natural gas industry — from the release of carbon into the atmosphere to the controversial technology of hydraulic fracturing (fracking). There are vast reserves of natural gas in North America that are trapped in shale formations. The natural gas companies have devised methods of releasing the natural gas from the shale by drilling into the shale and then injecting fracking compounds at high pressure to fracture the shale and release the gas. There are a number of environmental concerns with this process. First, the companies are not declaring what chemicals they are injecting, citing proprietary information and potential competitive harm. The wells cross through drinking water aquifers, and the fracking solutions have been found to contaminate ground water supplies. Companies generally create a pond of the fracking solution they extract from the ground. There are many reports of fracking operations causing gas seepage from the ground and into water supplies, sickening people and livestock.[22] The agricultural demand for natural gas is a component of the demand for natural gas. If agriculture switches to naturally fixed nitrogen, the demand for natural gas would decrease and the price of natural gas would decline, as would exploration activities.

A largely unseen impact of synthetic nitrogen is its pollution of nearby waterways and lakes. Nitrogen is one of the major contributing nutrients to algal blooms, phosphorus being the other. The release of nitrogen and phosphorus can cause a rapid overgrowth of algae and other plants, which overwhelm the ecosystem. This growth dramatically alters the water habitat and impacts the ability of other organisms within the ecosystem to survive. If the rapid algal growth is followed by a rapid die-off, significant concentrations of toxic chemicals can be released simultaneously with dramatic reductions in dissolved oxygen levels, further impacting aerobic plants and animals in the water.

The Great Lakes ecosystem, which contains 20% of the world's freshwater supply, is one of many ecosystems globally that is under extreme duress from excessive phosphorus and nitrogen loading. The

---

22. These reports are hotly contested by the natural gas industry; however, there are enough of them from across North America to have me believe that there is an issue. If you want the anti-fracking view, watch the movie *Gasland* (2010). It captures the stories from several areas where fracking is occurring.

National Wildlife Federation published a report entitled "Feast and Famine in the Great Lakes: How Nutrients and Invasive Species Interact to Overwhelm the Coasts and Starve Offshore Waters,"[23] which draws a connection between phosphorus and nitrogen loading and a 95% reduction in prey fish numbers in the space of 20 years. Additionally, toxic algal blooms are increasing in area and severity. The study examines the interaction of invasive species, including zebra mussels, the nutrient loading, and the complete collapse of the aquatic ecosystem in the lakes.

There is a distinct difference in the pollution potential between synthetic and natural source nitrogen fertilizers. Synthetic nitrogen washes off farms because the plant-available forms are water soluble, and for most crops, all the nitrogen needed is applied in the spring and then is slowly taken up by the plants over the course of the growing season. Natural source nitrogen is trapped in organic matter and is slowly released as the organic matter decays. The rate of release increases with increasing soil temperatures, which parallels the growing crop needs. This results in much lower levels of nitrogen leaving farms as pollution when natural source nitrogen is applied.

Conventional agriculture continues to bluff with the red herring that modern agriculture needs synthetic nitrogen fertilizers to maximize yield and therefore feed the world. What they're not telling you is that using synthetic nitrogen fertilizers also has a number of effects that lower the future potential for food production. If we switch from synthetic nitrogen to naturally fixed nitrogen, we simultaneously reduce greenhouse gas emissions, build soils, and reduce nitrogen runoff from fields.

But how do consumers know whether their choices are reducing our reliance on synthetic nitrogen fertilizers? There are two ways consumers can be assured they are making a wise choice. The first is to choose foods that have their nitrogen supplied using the natural process: legumes (peas, beans, lentils, chick peas, sea buckthorn). True grass-fed beef, lamb, milk, and milk products also derive all of their nitrogen needs from legumes in the pasture and hay. The second choice a consumer can make is to choose certified organic foods. The certified organic rules require farmers to use only natural nitrogen fixation — no synthetic nitrogen compounded using fossil fuels is allowed.

---

23. Julie Mida Hinderer et al., "Feast and Famine."

PHOSPHORUS

Life as we know it would largely cease without phosphorus. A phosphate molecule is the link between each of the nucleic acids on the DNA chain. Without phosphorus, the code of life literally falls apart. Phosphorus is also the key to the energy cycle that powers all aerobic organisms at the cellular level.[24] The Krebs Cycle is dependent on phosphorus-based compounds — adenosine triphosphate being the critical energy transport molecule. Phosphorus is also critical to all vertebrate and invertebrate life forms as it is one of two primary constituents, along with calcium, of bone and shell.

Phosphorus is mined in the water-insoluble form of phosphate rock. It is then chemically treated with acids to form various water-soluble salts, such as monoammonium phosphate (MAP), diammonium phosphate (DAP), and super phosphate, which are used as fertilizers in conventional agriculture. Organic agriculture uses ground-up phosphate rock directly.

There is a legend that phosphorus fertilization in agriculture was originally the product of the soap and detergent industries. Their manufacturing processes created a significant quantity of phosphorus by-products that needed to be disposed of. It's a great story, but it's not accurate. The process to derive super phosphate using an acid treatment was discovered in England by J.B. Lawes. His patent on the process in 1846 allowed him to accumulate substantial wealth, and he funded the trust that now operates the Rothampsted research farm, which has some of the longest-running fertility trials on earth. His process used bone as the starting point, but the process was later adapted to use rock phosphate as the starting substrate. In the mid 1800s, England was importing 40,000 tons of bone to be processed into phosphorus fertilizer in addition to the consumption of 26,000 tons of domestic bone. This predates the phosphorus by-products from the soap and detergent industry by many decades.

Most people are familiar with the concept of peak oil — the point at which our annual consumption exceeds our annual discovery of new reserves. Once we pass peak oil, the known reserves of oil will diminish and prices will rise due to the depletion of known reserves. There is a parallel concept for every nonrenewable resource on the planet. Some

---

24. There are a few metabolic pathways that aren't dependent on phosphorus — primarily in single-celled anaerobic organisms — but there is no known organism that doesn't use phosphorus in its DNA.

calculations suggest we passed peak phosphorus in 1989. Depending on who you believe, at current rates of consumption, known reserves will be depleted before the end of the 21st century or will last well into the 23rd century. As well, no significant exploration for phosphorus deposits has occurred since phosphorus has been readily available at low prices. It is likely that as supplies diminish and prices increase, new deposits will be found. However, our current use of phosphorus is not sustainable in the true long run. All that the different sources are arguing about is how long the depletion tail will last.

What's even more worrisome is that world phosphate production is concentrated in four countries: Morocco (and its neighbours in the Western Sahara), the United States, Russia, and China. Together these countries account for over 70% of world phosphate production. As for phosphate reserves, 77% of known reserves are in Morocco and the Western Sahara alone, a region that is volatile with borders in dispute. For point of comparison, the 12 member nations of OPEC produce around 40% of the world's oil and control about 80% of the world's known oil reserves.[25] We've had several oil-motivated wars: Will phosphate become the next mineral to generate conflict?

Certified organic agriculture is allowed to use the rock dust form of phosphate. It is essentially the mined phosphate mechanically ground into fine particles without using any acid or other chemical treatment. Thus, the depletion arguments are equally valid for both industrial and organic agriculture. From my perspective, phosphorus is the single most limiting nutrient in organic agriculture.

The issue of phosphorus has a couple of interesting conundrums. There are some soils that have built up excessively high levels of phosphorus. The excess phosphorus is a pollution problem. These soils are found around confined animal feeding operations. Animals consume feed and excrete part of the phosphorus in their manure. The manure is applied to the fields closest to the barns because it costs more to haul the manure than the value of the nutrients in it. Phosphorus builds up in the soil until it is at excessive levels. Holland had such a problem with phosphorus pollution that it issued phosphorus permits to farmers,

---

25. This number is supplied by OPEC. The percentage of reserves is not a hard number. For example, Canadian reserves are generally listed in the range of 178 billion barrels; however, there are estimates that include shale oil that put the number an order of magnitude higher at 2 trillion barrels. (A barrel is the standard volume measure for oil, equalling 42 US gallons or approximately 159 litres.)

limiting the amount of phosphorus they could spread. Legend has it that there was at least one marriage of convenience between someone who was generating more phosphorus than they had permits for and someone who had more permits than they needed. Many of the Dutch families now farming in Canada sold their manure permits and land in Holland for large sums and bought much larger farms here.

This is one of the unintended side effects of agricultural specialization and varying environmental constraints on agriculture. In the United States, for instance, trainloads of grain are leaving the Corn Belt to supply confined animal feeding operations in other states. These trains are not just hauling corn or soybeans; they are hauling the nutrients that comprise the soybeans and corn. A pollution problem is then created in the states where the animal feeding operations are because of the build up of the nutrients around the livestock operations. Meanwhile, back in the corn-producing region, mined fertilizers are being transported in by the trainload to replace the nutrients that were hauled away. Cheap fertilizer and cheap transportation have severed the complimentary relationship between crop and animal production.

OK, take a deep breath; I'm about to argue for the application of sewage sludge and dead bodies to cropland. A sustainable future depends on it. Unlike oil, phosphorus isn't "consumed" by use. It can be 100% recycled. In the natural cycle of an ecosystem, plant roots take up phosphorus and potassium from the soils, the plant is consumed by higher-order animals going up the food chain and it is eventually returned to the soils either as excrement and urine or through decomposition when the plant or animal dies. In the human food chain, the phosphorus and potassium get a one-way trip from farmers' fields to cities. Phosphorus also enters waterways through run-off and erosion from fields. More accurately, I'm about to argue for the recycling of the nutrients contained in sewage sludge and dead bodies for use on cropland. If we spread the sewage sludge, it is simply "reuse" of the phosphorus and nitrogen. If we further process the sludge to extract the phosphorus, then the proper term is "recycling."

Sewage sludge is nasty business, there's no two ways about it. The hysteria around sewage sludge being spread on farmland is largely nimbyism (a not-in-my-backyard attitude). A significant portion of the opposition originated with the broadcast spreading of the sludge and the associated smell. Most sludge is now injected directly into the soil and community opposition has been greatly diminished. The opposition now focuses on disease threats or heavy metal accumulation and other pollutants entering either our food system or our water supply. However, fertilizing with it is the most natural of all our processes for phosphorus fertilization. That's how it happens in the ecosystem. Animal crap is plant food.

I grew up in the shadow of a major metropolitan area, and many of my neighbours used sewage sludge as a source of nutrients. I probably would have used it as well, had I not switched to certified organic production. It was/is the cheapest source of nitrogen and phosphorus on the market — it's free. The problem with sewage sludge is that it isn't solely human excrement. People and industry like to flush nasty compounds into the sewers. A portion of the sewage sludge is incinerated, a portion is "recycled" onto land, and a portion is deposited in landfills. Approximately 50% of North American sewage sludge (also known as biosolids) is spread on farmland. Certified organic rules prohibit the use of sewage sludge.

Unfortunately, we consume more than just natural foods and we dump non-biological waste down our drains and in our toilets. The drugs we take are excreted by our bodies. The cleaners and detergents we use flow into the sewers. The old medicine we dump in the toilet ends up in the sewers. If we are sick, the disease-causing organisms end up in our sewers. Even household insecticides and rat poison end up in the sewers.

To recycle the phosphorus in sewage sludge, the question becomes how do we limit the risks from all the other substances it contains? Most jurisdictions in the developed world have regulations composed of three elements for spreading sewage sludge: set levels for known pollutants and monitor the sludge, limit how much and how often sludge can be spread on any particular field, and regulate the location of sludge spreading to minimize the chance of contaminating waterways or inadvertent human contact. For example, there has to be a setback from property boundaries, watercourses, and wells to minimize the risk of the sludge moving off of the field or someone unknowingly coming in contact with it. There is also a setback in time for crops that touch the soil or are grown in it. For example, a farmer can't apply sludge and grow potatoes or carrots in the same year or the following year in most jurisdictions.

I tend to agree with the concerns about the unknown compounds in the sludge. The technology exists to remove the heavy metals during the sewage treatment process. Some municipalities use the technology, but a good percentage does not. It's the same with the pathogens in the sludge; technology exists to sanitize the sludge, but sewage treatment plants are not required to use it. Somehow we have to find acceptable ways to recycle the life-giving elements within the sludge without spreading the pollution with it. The sooner we figure out an acceptable recycling strategy, the better. This recycling process will shrink the waste heading for landfills and will limit the necessity to mine our landfills for phosphorus in the future.

Even if we collectively decide that the risks from sewage sludge are greater than the benefits, there are laboratory-scale technologies for extracting the phosphorus from sewage sludge and separating it from the heavy metal content. Some methods are energy positive in that the energy created through the process is more than is needed to extract the phosphorus, and the surplus could be used to generate electricity. However, at the long-term average phosphorus price, most of these processes are not profitable.

So how do we encourage phosphorus recycling and divert it from landfills and waterways? We can continue to expand sewage sludge spreading on farmland. I've already registered my uneasiness with this solution. Some municipalities are composting the sludge before application. I personally think that any composting process that involves the addition of fossil fuels to the organic matter (through the engines of the equipment turning the compost) is completely missing the point of composting. Plus, the composting process doesn't remove the heavy metal content of the sludge; it only reduces volume, eliminates odour, and decreases the pathogenic organism content if the composting is done correctly.

There is a way to encourage phosphorus recycling without intervening in the marketplace. If the pure form of phosphates that are recovered from sludge through the extraction processes were added to the approved list for use in organic agriculture, a market would be created that could sustain a premium price regardless of what happens to commodity phosphorus prices. Purists will say absolutely not. Phosphorus recovered from sewage sludge is a completely man-made, synthetic product. It is the antithesis of what organic stands for. I take a slightly different view. By recycling the phosphorus content of sewage sludge, we are moving towards a sustainable future. We are reducing the phosphorus loading of our waterways and the associated ecological damage. We are improving the productivity of organic farms by providing an available source of the primary limiting macronutrient, thereby improving the energy balance of organic farms. From my perspective, it's a win-win-win situation — humanity benefits, the environment benefits, and the most valuable resource is conserved.[26]

Chinese agriculture was sustained for four millennia through an obsession with returning all organic matter back to the fields. It was considered impolite not to go to the bathroom after consuming a meal at someone's house. The survival of humanity will ultimately depend on developing a similar obsession with recycling phosphorus. That may sound melodramatic, but of all the threats to our food system, the depletion of phosphorus reserves is the one problem for which there isn't an alternative solution. We have to learn to recycle phosphorus or face a permanently diminished capacity to produce food. If our society wasn't so indiscriminate with what was flushed into our sewer systems, we wouldn't even be discussing this issue. By not recycling phosphorus we are both diminishing our capacity to produce food in the future on land and impacting our ability to harvest food from our aquatic resources.

How can a consumer's dietary choices reduce phosphorus pollution and improve the recycling of phosphorus? Switching to organic food can reduce phosphorus pollution. The MacRae et al literature review concluded that there was "solid" evidence that organic farming systems reduce soil erosion and phosphorus loading from agricultural activities.[27] The 30-year Rodale Farming Systems study reached similar conclusions.[28] However, the organic rules lock us into consuming phosphorus without recycling, whereas conventional agriculture has the flexibility to recycle phosphorus.

The biggest contribution you can make to the recycling of phosphorus is to stop putting anything down your drain/toilet that isn't naturally a plant food. Clean your toilet with vinegar rather than a synthetic chemical cleaning product — we do, and it works. (You have to let it soak a little longer than the strong stuff, but vinegar is acetic acid — a natural acid formed by fermenting sugars past the alcohol stage). Or you can switch to biodegradable cleaning products. One of the steps in a

---

26. I hope I don't have to repeat myself. Without phosphorus, life as we know it ceases to exist. Of the life-critical resources, phosphorus is the only one that is currently managed as a straight depletion with a known end point.
27. See note 14.
28. Rodale Institute, "Farming Systems Trial."

sewage treatment plant is digestion by bacteria. And if you happen to be involved in the organic movement, encourage discussion of a phosphorus-recycling program. Many people I have raised sludge spreading with haven't considered the recycling of phosphorus, only the prevention of other compounds contaminating farmland.

POTASSIUM

Potassium is the third macronutrient typically applied in fertilizer. While potassium isn't incorporated into the structure of living things in the way that nitrogen and phosphorus are, it is critical to most of the life-giving functions within the plant. Potassium is:

- a catalyst in photosynthesis;
- a catalyst in reading DNA and forming proteins;
- critical to the transport of sugars throughout the plant;
- critical to the absorption and transport of water and all water-soluble nutrients within all life forms;
- a catalyst in the formation of starch for energy storage; and
- critical to the mechanism by which plants conserve water during a drought.

The primary source of potassium for fertilizer is mined potash (KCl). Over 50% of the world's known reserves are in Saskatchewan, Canada. At this point, known reserves are growing faster than potash is being mined. There are currently no estimates of "peak potash." The environmental impact of potash and the allowable forms in organic agriculture have been debated since the beginning of organic agriculture. The fact that potash is almost half chlorine by weight and that potassium chloride has a high salt index has made biological farmers wary of using potash. A similar compound in terms of salt index is common table salt (sodium chloride [NaCl]). There are many legendary historical references to salting the fields of conquered enemies to destroy the enemy's ability to grow food and rebuild their civilizations. It is true that excessive levels of salt in the soil will kill the soil ecosystem and render it unusable for crop production. However, the historically high price of salt makes the plausibility of the claims questionable.

57

Originally, many organic standards prohibited the use of potash, but the current national organic standards allow natural potash to be spread on organic fields. The only difference between the natural potash and the potash used by conventional agriculture is that the conventional potash has a small amount of oil added as a dust suppressant and some forms of potash have been adulterated to ensure a uniform particle size and therefore uniform spreading.

So far, potassium has not been identified as a pollutant of concern. Neither Canada nor the United States has established regulations governing potassium levels in water. At this time, an eater shouldn't be concerned about the effect of their dietary choices on potassium use. Potassium levels are very low in sewage sludge and therefore the recycling arguments that apply to phosphorus do not apply to potassium.

OTHER MAJOR NUTRIENTS

The other macronutrients needed for life — carbon, hydrogen, oxygen and calcium — are available to plants directly from the air (in the case of first three) or are abundantly available in the soil (in the case of calcium). Nutrients that have a gaseous phase can be recycled back to the farms without intervention by humans. Nitrogen, carbon, hydrogen, and oxygen have no trouble escaping the cities and returning to the countryside to be reused in the formation of more food, although recycling urine would allow nitrogen to be returned without having to go through the energy-intensive fixing process.

Unfortunately, the same cannot be said for phosphorus or potassium. They are on a one-way trip from mine to farm to buffet to toilet to landfill or the ocean. Our known potash reserves will sustain us well into the twenty-third or twenty-fourth centuries. The first challenge will be phosphorus.

Changes to your diet and substituting synthetic chemical cleaning products with biodegradable ones will have an impact on the fertilizer available to future generations to grow their food.

CHAPTER 5

PESTICIDE: POISON OR ANTIDOTE?

"All things are poison, and nothing is without poison; only the dose permits something not to be poisonous."

Or more commonly,

"The dose makes the poison."

— *Paracelsus*

The first pesticide I ever applied was malathion, an organophosphate insecticide. I was somewhere around ten years old. Dad would pull into the yard with a truck of wheat during harvest, set it up to unload, and then have me watch it while he grabbed some lunch or did an odd job. My role had three main tasks. The first was to make sure that the wheat was flowing out of the truck without it overloading the auger, which was taking the wheat up into the storage bin. The second was to shut everything down quickly if something went wrong. The third job was to sprinkle malathion onto the wheat as it flowed into the auger. I had a 25-kilogram bag of malathion beside me and a 1.5-litre juice can with which I scooped the malathion out of the bag and sprinkled it onto the wheat. I was bare handed and wasn't wearing a dust mask. We were adding the malathion to the wheat to prevent against an insect called a grain weevil. Their larvae eat the insides out of kernels of grain, they multiply quickly, and no grain elevator or miller will buy grain contaminated with them for fear of their wheat storage becoming infested (rightly so).

Adding malathion was a common practice in those days, but today, improved herbicides and grain storage bins have substantially reduced its use. Improved herbicides mean much less green weed material with the potential to cause damp hotspots where weevils thrive ends up in the bin.

Today's grain bins are also more effective at drying the green stuff and cooling the field heat from grain. Every bin I've built has had full floor aeration — there is a false perforated floor in the bin and high volumes of air are blown in under the floor. The air travels up through the grain, removing moisture and heat as it goes, and then exits the bin through vents in the roof. By having well-designed storage facilities and the ability to prevent hotspots, I have so far eliminated problems with weevils.

I shudder as I think back to my young self sprinkling that malathion onto the wheat. At the time, I was proud of myself for having grown up enough to be left to watch over part of the harvest. I knew how to raise the dump box on the truck, I knew how to start and stop the tractor, I could open and close the chute on the truck, and I could start and stop the dump truck. I was in the big leagues. We knew malathion was a poison — there was a skull and crossbones on the label. I was warned to stay out of the dust and to try not to inhale any, but we didn't take any of the precautions I would automatically take today, like chemical-resistant gloves, filtered dust masks, and eye protection.

My first introduction to proper safety protocols came in the summer after my first year of university, when I worked as a summer research assistant for a major agricultural chemical company. We had to follow strict safety protocols when working with both registered and experimental chemistries. One portion of their business was in corn rootworm insecticides. They had the market-leading insecticide but were researching ways to reduce the dust in their formulation. They had a couple of promising formulations that were ready for field-scale trials, so I volunteered Dad to be one of the farm co-operators. Dad was a little amused when we came out to set up the trial as he was planting — we had on full moon suits, chemical-resistant gloves, eye protection, and filtered masks. He was even more amused when we came back in August to do rootworm counts and were again suited up before digging up our soil samples from the root zone. It was a very different paradigm with respect to the hazard of the pesticide.

By the time I graduated from university, the province had brought in a mandatory licensing system for pesticide applicators, and every farmer had to attend a one-day course and pass an exam to be licensed. A good portion of that course was focused on safety for both the operator and the environment. I ended up teaching the course for a few years as a source of extra income in the winter. It was amazing how much the industry had changed in the decade between me being a kid trickling malathion into the grain and me becoming an instructor for the Ontario Pesticide Education Program.

Now, before you call Children's Aid and inquire about having my father charged retroactively, let's put that malathion application in context. The lethal dose for malathion is three times larger than the lethal dose for caffeine, and you get far more caffeine in a cup of coffee than I would have absorbed sprinkling the malathion onto the grain. Does that mean malathion can be used indiscriminately? No, and I'm sure my liver was working overtime many of those days as a child. Our perceptions of safety and risk evolve over time. I was sprinkling malathion without safety protection during the same era when mandatory seatbelt laws were gradually being enacted across North America. Do you remember the opposition to the "nanny state" interfering in our lives? Yet today, most of us wouldn't even consider driving down the highway without our seatbelts on.

NATURAL AND SYNTHETIC PESTICIDES

Warfarin is an example of a synthetic compound derived from a natural source that is toxic at some doses and life saving at others. It is a classic example of "the dose makes the poison." Those of you who are at risk of pulmonary embolisms will recognize warfarin as a "blood thinning" drug that is dramatically extending your life. Those of you who have had problems with rodents will recognize warfarin as rat poison. And those of you who remember when sweet clover was grown more commonly will recognize bicoumarin (from which warfarin is derived) as the toxin responsible for sweet clover bleeding disease in animals that eat mouldy sweet clover hay. Warfarin can kill a horse or save a human life depending on the dose. The horse would have been killed by a naturally occurring toxin, and the human's life would be saved by a synthetic medicine.

I'm afraid the pioneers of the organic movement may have clouded the discussion of pesticides permanently by making a distinction between naturally occurring and synthetic pesticides. The list of deadliest known poisons, for example, includes many naturally occurring compounds — asp venom, skin secretions from the poison dart frog, spider venom, and curare, to name a few. In farming, we have a class of insecticides called pyrethrums: some are synthetically compounded, and some are extracts of the pyrethrum daisy. The latter can be used in certified organic production; the former are prohibited substances. The natural versus synthetic distinction made in the organic rules isn't about toxicity. Conventional agriculture apologists like the toxicity argument because organic has drawn some arbitrary lines that are hard to defend from a strictly toxicological point of view.

The organic reasoning for these distinctions is that anything that comes from nature can return to nature, and there will be a pathway for it to decompose into harmless substances that become food for another organism. Our bodies also have several pathways for removing toxins — the liver, kidneys, and to a minor extent, the skin through sweating. We willingly ingest foods containing natural toxins at almost every meal. If we were to exclude everything with a measureable level of a known toxin from our diets, most people would be surprised by the foods that would disappear. The following list explores foods containing known toxins, but it is by no means complete:

- All solanaceous crops (members of the nightshade family), including tomatoes, peppers, eggplant, and potatoes, contain a toxic glycoalkaloid in at least part of the plant. The best known of these is solanine, which is the toxin found in the green skin on a potato that has been exposed to light. Similar compounds are also present in the green, unripe fruit of other members of the nightshade family.

- Almond extract and amaretto are derived from bitter almonds, which have high levels of prussic acid. Prussic acid is converted to the highly toxic hydrogen cyanide gas during digestion. Consuming a handful of bitter almonds could be fatal; however, there is no evidence that the prussic acid is present in the almond extract or amaretto.

- Prunes, plums, cherries, and apricots are all members of the same botanical family as almonds and have detectable levels of prunasin (a cyanogenic glycoside), which is also converted to cyanide gas during digestion.

- Cassava (also known as tapioca, or manioc) in its raw form, cassava root, is poisonous due to the presence of linamarin, a toxin that is chemically similar to sugars but has a cyanide ion attached. Linamarin is broken down in the digestive system, releasing cyanide gas. Eating two raw cassava roots is sufficient to cause fatal cyanide poisoning. Some researchers are looking at GMO methods to reduce the levels of linamarin in cassava root.

- Raw soybeans contain a trypsin inhibitor. Trypsin is a key enzyme in the small intestine that facilitates the digestion and absorption of protein.
- Most grains and legumes contain some form of lectin, a group of biologically active glycoproteins. The primary toxic impacts of lectins are the grouping of red blood cells into masses and irritation of the lining of the stomach and intestines to the point of causing bleeding lesions. Castor beans, containing a lectin called ricin, are the most extreme. Ricin is a two-part holotoxin molecule with a lectin (Ricin A) bonded through a single covalent bond to a second similar compound (Ricin B). When the two are joined, the toxicity is greatly increased. Ricin A is found in many grains including barley. The lectin level in black beans will kill a rat after a steady diet for a week. However, not all lectins are toxic (one lectin is being researched as a partial treatment for HIV), and lectins can be deactivated through thorough cooking.

Chronic toxicity is another dimension of the detrimental impact that chemicals (natural or synthetic) can have on humans and the environment. Chronic toxicity refers to a group of impacts that can occur from repeated exposure to a chemical. The major categories are:

- carcinogenicity – causing cancerous growths;
- mutagenicity – causing mutations to the genetic code;
- teratogenicity – causing birth defects;
- oncogenicity – causing tumour growth (not necessarily cancerous);
- organ damage – causing damage to internal organs, such as cirrhosis of the liver caused by chronic alcohol consumption;
- reproductive disorders – causing a reduced ability to reproduce, such as lowering sperm count and sterility, or miscarriage;
- nerve damage – irreversible damage to the nervous system, such as the cumulative impact of repeated exposure to organophosphate insecticides on cholinesterase depression; and
- allergenic sensitization – the development of allergic responses through repeated exposure.

Whether a chemical is man made or natural has very little bearing on whether it has any of these impacts.

63

Pesticide residues are routinely being found on fruits and vegetables (albeit at levels well below the safe levels established by regulators). Is that a problem? Maybe. Roundup is one of the most widely used, discussed, and disparaged pesticides on the market today. You might already be concerned about whether you're getting a dose of it in your food. Did you know that it takes 20 times more Roundup to kill you than caffeine? If you are consuming any as a residue on the food in your diet, you can be sure that it is significantly less than the amount of caffeine most people consume daily. All pesticides have an established safe residue limit based on acute toxicity research. We have the ability to detect pesticide residues at ever decreasing concentrations. Whether any of these residues have an impact on human health is debatable. Certainly on an acute toxicity basis, there is very little to worry about; however, chronic exposure beginning in childhood is a reality that we have collected very little research on.

But then, we haven't collected much research on the impacts of all the natural toxins in our food either. We started eating some foods when the only real understanding of toxicity was binary — either you felt sick when you ate something or you didn't. Can you imagine being the first person that said, "Hey, I think if we grate these poisonous cassava roots, soak them for 24 hours, and then cook them, they'll be safe." Or how about the first guy that decided to eat an almond. Every other pit of stone fruits is toxic, and bitter almonds are poisonous. In the case of almonds, you can imagine that they observed birds and small mammals eating the pits (an almond is not technically a nut) and decided that it was safe to try. But what about the cassava? There was no animal soaking it for 24 hours and cooking it. You have to wonder how that got figured out. The same thing goes for beans. Eat them raw and they're toxic; soak and cook them and you're OK.

Does the fact that we eat foods with toxins in them mean we should ipso facto tolerate additional toxins being added to our food? No. However, just because something is toxic at some level doesn't mean that we should automatically exclude it just because we can detect it. By that logic, we would have very few foods left — we would likely be reduced to a largely carnivorous existence of raw meat, since cooking over a flame introduces toxic and carcinogenic substances into meat (sorry barbeque lovers). Oh wait, what about the possibility of bacterial contamination on the meat? Maybe we should reduce the risk of food poisoning by cooking and risk ingesting the small concentrations of carcinogens that potentially contaminate the cooked meat. We accept trade-offs. The benefit of food safety and that great smoked barbecue flavour outweighs the risk of the addition of carcinogens for a majority of the population.

ENDOCRINE DISRUPTION

There is an emerging line of research looking at the phenomenon of endocrine disruption caused by very low exposure to chemicals. The endocrine system of the body is a finely tuned mix of hormones and receptors that control the important systems in the body. Oestrogens, testosterone, serotonin, dopamine, and epinephrine are all examples of compounds that are integral components of our endocrine systems. The functions regulated by the endocrine system include formation and functioning of the reproductive organs, the development of babies in the womb, brain development, and musculoskeletal development. Dr. Theo Colborn, a leading researcher in this field, and other researchers from around the world had accumulated enough evidence by the early 1990s that they authored a book, *Our Stolen Future*, published in 1996. The book describes the evidence and reaches a convincing conclusion that endocrine disruption is a real and dangerous phenomenon.

What's been done about this in the last decade and a half? Nothing. Well, virtually nothing. Bisphenol A (BPA), an endocrine disrupter in plastics, has made the news and has been largely eliminated from food containers because of endocrine disrupting properties. Dr. Colborn has also accumulated a database of 800 compounds that indicate endocrine disrupting properties according to documented, peer-reviewed citations. Three chemicals on that list are glyphosate, caffeine, and atrazine.

Have pesticide registration requirements changed? No. Why? It's complicated. The pesticide industry is terrified of endocrine disruption. The entire registration and regulatory framework for pesticides is underpinned by a fundamental assumption that the smaller the dose of poison, the less toxic it is. Due to this assumption, research is structured to discover the lowest dose that has an adverse effect and then the safe level is set at a dose that is many multiples smaller. If the pesticide industry accepts endocrine disruption as a valid impact, there is no linear relationship between toxicity and dose and no short cut to determining safe levels. Endocrine disruption can occur at doses that are several orders of magnitude lower than currently established safe levels.

Instead of embracing this line of research, the pesticide lobby has very quietly continued to spread doubt about endocrine disrupting impacts. They have also failed to research the endocrine impacts of their pesticides. I expect the reason for this relates to tort law — once it can be proven that a company knew about a potential detrimental impact, that company can be sued for negligence. The easiest way to avoid knowing about a negative impact is to not research it. As long as the industry can stall new regulations, they can continue to sell their products.

There is also an evidentiary problem with our current regulatory framework. Once a pesticide is registered, the burden of proof shifts from the manufacturer to society. Until society proves that it is unsafe, sales continue. Even if new evidence surfaces that points to a substance being unsafe, the burden of proof doesn't shift to the manufacturers to disprove the concerns until there is a significant accumulation of evidence that triggers a registration review. Fortunately, both Canada and the US are slowly working through an evaluation of all registered pesticides to ensure that the compounds would meet current standards. The reviews have resulted in many deregistrations, including almost all organochlorine-based insecticides.

Endocrine disruption detractors like to cause doubt by asking where all the impacts of these chemicals are. If they're so dangerous, shouldn't all of humanity be depressed, deformed, and have dysfunctional bodily systems? In other words, should we really be worried about endocrine disruption? Scientists are compiling compelling evidence that suggests many of our modern plagues have their roots in endocrine disruption: rising diabetes rates, increasing birth defects, increasing reproductive problems, increasing rates of depression, and increasing rates of ADHD. Einstein famously said, "The significant problems we face cannot be solved by the same level of thinking that created them." There's an ironic twist to endocrine disruption. Evidence suggests that endocrine disruption has the potential to lower the average intelligence of society.[29] In another generation or two, we may not be able to muster intelligence greater than the intelligence that caused our problems.

29. Theo Colborn, "Neurodevelopment."

Most of you will remember the kerfuffle that was raised around BPA. A particular focus was placed on baby bottles and soothers. The primary concern was that BPA would leach out of the plastic, into the milk, and be consumed by a baby. However, if that bottle contained soymilk, there was a much greater risk to the child due to the endocrine disrupters in the soy. Soybeans and other legumes have worrisome levels of plant estrogens, which have been shown to disrupt the role of natural estrogen in humans. If mothers were truly worried about endocrine disruption, they should have been throwing away the contents of the bottle before the bottle itself. As a final note, "BPA free" does not mean "endocrine disruptor free." Several of the replacements for BPA in plastics aren't any safer, they just aren't as well known.

I hope you've seen from the preceding analysis that we are confronted with many risks and trade-offs. Even without the man-made chemicals, we have toxins and potential carcinogens in our diet and the environment. The question becomes, is the risk from adding pesticides to our environment and food supply worth the benefit? If you're in conventional agriculture, the answer is likely, "Yes. The perceived increases in food production achieved with the use of pesticides are fundamental to feeding the world." However, I'm not sure that a full risk–benefit analysis, if honestly completed, would come to the same conclusion. The hearty "Yes" from conventional agriculture also assumes that food production would be lower in the absence of pesticides.

Even if you set aside the discussion of whether land-based food production is higher with the use of pesticides, you still have to account for the impact of pesticides on the aquatic environment. Around a billion people globally rely on fish as their primary protein source. We have irrefutable evidence that the fisheries that haven't already collapsed are about to. The amount of pollution we have dumped in the oceans has surpassed the oceans' ability to cleanse themselves and produce the protein harvest we are demanding. Are pesticides solely to blame? No. But pesticide and fertilizer runoff from agriculture make a significant contribution. The dead zone in the Gulf of Mexico is partially attributable to the soil, fertilizers, and pesticides flowing down the Mississippi River every year. This is no different from the dead zones at the mouths of every other major river system in the world.

THE INTERPLAY BETWEEN PESTICIDES AND OTHER DIMENSIONS OF SUSTAINABILITY

Given that there is the potential to produce equivalent yields without using pesticides and that pesticides negatively impact our ability to harvest food from lakes and oceans, why haven't we banned the use of all pesticides? Unfortunately, it's not that simple. My experiences from my early years in organic production help to demonstrate this.

From 2005 to 2008, my partner and I ran 2000 certified organic acres. However, we didn't have sufficient livestock to go with that many acres. We tried substituting with purchased compost and rock phosphate dust, but without the livestock to consume the hay from the legume portion of our rotation, we were always minimizing legumes. We kept them in the rotation a minimum of once every five years to meet the organic standard; however, it was never for three consecutive years. It was always a one-year crop of red clover or sweet clover that was ploughed in. The difference in the impact on soil, weed pressure, and pests is substantial. When you leave soil undisturbed for several consecutive years, you let the soil life stabilize. The soil aggregates reform. Worms have a chance to digest residue from the surface and concentrate phosphates. Annual weeds disappear and perennial weeds are severely weakened.

Why did we take this shortcut? Lack of knowledge and greed. First off, we were 80 miles from one corner of our land base to the other. It's one thing to harvest grain and put the yield from 20 or 40 acres in a truck and haul it home. It's quite another thing to harvest any quantity of hay and get it hauled any distance before it spoils or gets rained on. I viewed the legume requirement as a necessary evil rather than a critical part of the system I was managing. I was still partially in my conventional mindset. Each problem had a distinct cause and a distinct solution. That solution could be in the form of an input, timing of tillage, or a specific tillage tool. I didn't look to the design of the system to correct the problems.

Depending on how you view the relative importance of erosion versus pesticides, it could be easily argued that I abused land worse in my beginning years in organic production than in my conventional years. Before I switched to organic, I had put the mouldboard plough away and was using almost 100% conservation tillage and no-till. That means that I tried to disturb the soil as little as possible between crops to reduce the amount of erosion. The mouldboard plough is a fantastic but evil tool. It takes the top 10 to 15 centimetres of the soil profile and turns it upside down. It's a great tool for "burying your mistakes"; however, it leaves the soil completely bare and at severe risk of erosion.

In conservation tillage, you are relying on herbicides rather than tillage to kill perennial weeds. When I switched to organic, I tried to continue with conservation tillage, but gradually the perennial weeds became a problem. The neighbours joked about my "crop circles." We have Canada Thistle as a background weed. One plant establishes and then over the course of a couple of years, the rhizomes grow outwards and you have a circular patch of thistles in your field that chokes out anything you plant. The organic thinking at that time was that you mowed the thistles just before they flowered to weaken the plant and prevent seed from forming. I had a particularly bad field right beside one of the main highways in the county. After I mowed all the thistles, it looked like I had a number of crop circles. We never did get them weakened enough to kill them.

We ended up summer fallowing that field for an entire season and then planting winter spelt. Summer fallowing means that I mouldboard ploughed the field in the fall. The bare soil was left exposed to the elements for the entire winter. All the next spring and summer we cultivated the field every two weeks to kill anything that had grown. By the next fall, we had killed the thistle, but the soil was like sifted flour rather than crumbly dirt. We had destroyed the soil structure. (Our neighbours even complained about the dust blowing.) So I ask you: Was that better than spraying two litres of Roundup? I can't see how.

Now that my farm is animal based, thistle problems are dealt with in a hay field. We try to cut hay three times a year. Each time, the thistle gets cut off and the alfalfa and grass quickly grows to compete with it. Over the course of several years, the thistles are weakened and die. Is that better than two litres of Roundup? I'd say absolutely.

To achieve organic yields that match conventional ones, the farm has to be managed as a complete system. A following chapter gets into the details and the practicality of a farming system that can achieve equivalent yields to conventional without the use of pesticides. It's experiences in my past like this that lead me to say I'm a reforming industrial farmer. I'm still learning about the new system. The organic production manuals are a starting point, but they're not the end point. It takes more than just the organic rules to convert a farm to a sustainable alternative to conventional agriculture, and the following chapters will lay out the necessary steps as I see them.

THE HERBICIDE CHOICE CONUNDRUM

Before I leave the issue of pesticides, I want to lay out some of the choices that conventional farmers are making. I truly believed that I was improving our farm's environmental footprint when I was a conventional farmer. I think the research would bear me out on that fact. My first years in organic production were more environmentally damaging than the conventional operation that I transitioned to organic production. But the environmental damage was in a different form.

In my mind, you can't talk about Roundup in a vacuum. It has to be put in the context of the alternatives. Most of you have likely heard of Roundup or possibly even used some in your yard, but very few of you will have heard of the herbicide atrazine. I've already told you about summer fallow and my pasture/hayfield methods used to control weeds. Before there was Roundup, the availability of atrazine was single-handedly the most important technology in the growth of corn production acres. Atrazine kills just about everything but corn. If you put on enough, it will even kill nasty perennial weeds such as quack grass and thistles. Unfortunately, there is a catch. Atrazine is persistent. If you use the application rate needed to kill perennials, you are stuck growing corn in that field for three consecutive years because no other crop will survive. Corn after corn after corn leads to severe problems with rootworm, which means a strong insecticide is needed in the second and third years of corn in that field.

When I was growing up, every corn planter had a box for seed and a box for insecticide on each row. The insecticides had to be banded over the seed to protect the developing roots. I can remember dad filling the insecticide hoppers by pouring the granular insecticide bare handed from a paper bag into the hopper while trying to avoid inhaling the inevitable cloud of dust that arose. By the time I was taking over from dad, I was wearing chemical-resistant nitrile gloves and had on a cartridge mask that filtered out dust and organophosphates. That was the price you paid for ridding your fields of perennial weeds for a few years. Oh, and because atrazine was persistent and water soluble, it found its way into ground water and began showing up in rural wells.[30]

Along came Roundup, now the most popular herbicide in North America. It was revolutionary — we went from applying 2 kilograms of active ingredient and being forced to plant corn for a minimum of three years to applying 300 to 700 grams of active ingredient and being able to plant any crop we wanted. Rotations quickly adjusted to a single year of corn in rotation with soybeans and wheat. The nasty insecticides were no longer needed, and the insecticide boxes disappeared from corn planters.

As well, Roundup was the watershed herbicide for the increased adoption of reduced tillage and no-till. In no-till, the planter opens up a slot in untilled ground, places the seed, and closes the slot. Since you aren't doing any tillage, weed control has to be by herbicide. Erosion rates decrease significantly, and the residue from the previous crop is left on the soil surface to form mulch and gradually be digested by the soil life. That mulch also protects the soil from the impact of heavy rains and significantly reduces erosion.

On the surface, Roundup is a soil- and water-conserving herbicide. But now, Roundup is used so ubiquitously that the US Geological Survey has consistently measured it in the air and rain in major agricultural regions.[31] We're not sure what the consequences of that are, yet. With this in mind, I remember attending a lecture of a retired crop science professor while I was at Guelph. In that lecture, he declared that atrazine was the greatest thing that ever happened for corn production and that he had little time for its detractors. This was at a point in time when we already knew atrazine was showing up in ground water and that it disrupted the sexual development of amphibians. A decade later, we would learn that the impact on amphibians could occur at levels 30 times less than what the EPA considered the "safe" level of atrazine in drinking water. Like atrazine before it, I suspect we are on the cusp of research that shows Roundup isn't the innocuous herbicide we have been told either.

At the other end of the safety spectrum, environmental movements tend to focus on cause célèbres. I understand why. Focusing on a sharp point that can be easily communicated is a proven strategy for both fundraising

---

30. Atrazine is still the second most used herbicide in North America. It is used at lower rates that allow crops other than corn to be planted the next year, but it is still a very effective tank mix partner to give other herbicides an extra jolt.
31. Feng-chih Chang, "Occurrence."

and building momentum. But issues are never as simple as the sharp point. If Roundup and all its glyphosate, white-label competitors are banned, what will step in to replace them? What will be the impact on the environment and the acres of soil conserving no-till? I guarantee there won't be one big come-to-Jesus meeting and mass conversion to organic and grazing-based methods, but they will be replaced by something. I can't predict whether that will be good or bad. So the question becomes, what are the trade-offs we are willing to accept? Organic production, on the one hand, analyzes risk using the precautionary principle, which precludes the use of anything with a known potential risk, even if the risk hasn't been scientifically proven. From a conventional agriculture perspective, the benefit of weed or insect control and increased yield is a reasonable trade-off for the small risk of the added toxins to the food supply. However, with research showing that organic methods can produce equivalent or higher yields, the question becomes what benefit are we receiving for the increased toxin load in our food supply? From my perspective, there is none. The challenge we face is that our entire agriculture research and university and college education systems receive significant funding from the pesticide, fertilizer, and biotechnology companies. Organic systems research has a very small chance of succeeding in that environment. It amazes me that so many long-term systems trials in organic production have been able to succeed and that they were properly managed to produce credible results.

We also have to look beyond just the risk to eaters. Farmers, farm workers, and chemical industry workers all have a much higher risk of exposure than eaters. The 1984 Bhopal disaster in India was a leak of one of the compounds involved in manufacturing insecticides. Over 2000 people were killed and half a million were injured. When we choose to use pesticides, we are choosing all of the injuries and accidents that occur along the supply chain.

As a consumer, what is the easiest way to reduce your risk from pesticide exposure? Stop buying them. All of the ant and bug killers and lawn weed and feed products that you have in your house are a far larger source of pesticide exposure than your food supply. However, if you want to reduce the acres of land covered in pesticides every year, the best way to accomplish that is to switch to organic foods. But you also have to consider the previous chapters on erosion and fertility. Simply switching to organic isn't enough to encourage a truly sustainable system of food production.

You can choose to eat cheaper, conventional produce, but you are risking unknown damage to yourself and the environment. You would be choosing the unsafe work conditions for farm workers handling the pesticides, the working conditions in the pesticide factories, the risk of the environmental damage from spills, and the environmental damage where the pesticides are produced. You can choose.

CHAPTER 6

IMB, GMO, EIEIO

The saddest aspect of life right now is that science gathers knowledge faster than society gathers wisdom.

— *Isaac Asimov*

I own a few of one of the rarest and purest breeds of cattle on the planet — White Park. DNA testing has shown that they have the most distinct set of genes of all the breeds of cattle descended from the domestication of Bos taurus. All the other breeds are more closely related to each other than they are to White Parks.

The White Parks are my "hobby." We raise them alongside our herd of commercial cattle. My main purpose in raising them is not to produce beef — yet. The White Park is a legacy breed that was on the verge of extinction, with less than 500 animals left. They have such a distinct set of genes that there is likely something of value in their genes that can't be found in any other breed. The cattle industry just doesn't know it needs that gene — yet.

Physically, White Parks have two features that distinguish them — they have lyre-shaped horns and are white with black points (ears, socks, nose, and generally teats and tongues). The genes for their markings are so pure and dominant that if you cross-breed a White Park with any other breed, the resulting calf will be white with black points. The calf may take on other features of the other breed — size, stature, lack of horns, or muscling — but the colouring will stay true. I have some calves that are one-quarter White Park that still look pure.

There are a few genes for other colourings sprinkled throughout the population, but they are rare. Red points rather than black is the most common of the alternative colourings. A separate sub-breed has been established called "Chillingham" after the Chillingham Estate in England where the herd had predominantly red points. There is a much rarer gene that gives a solid black animal. This colouring is so off-type that they can't be registered as purebred White Parks. Until last year those were believed to be the only alternative colours in the White Park gene pool, but the farm in Virginia from which I purchased a new bull had a solid red calf last year. This is the rarest gene combination — no one in North America had ever seen this from purebred White Parks before.

If I were so inclined, I could select animals from within the White Park breed and end up with a herd with red points or that are solid black or solid red without crossbreeding with any other breed of cattle. I would start with a group of cows and bulls that are from lines known to throw the odd black calf, for example, and keep breeding within that pool and selecting black animals until I had a group of black animals that only produced black calves. However, if I wanted to create a "White Park" with any other markings, I would have to cross it with another breed to add the genes for the required markings and then select daughters and sons that had those desired markings. This is essentially the process that traditional breeding follows. Breeders keep selecting for a particular trait or traits until they've largely eliminated the genes that express variations of that trait in an animal or crop.

The challenge with traditional breeding techniques is that the diversity of genes gets narrower with each generation. As the base gets narrower, the room for improvement also narrows. You can outcross with wild relatives to try to introduce new genes and then select for superior traits, but it becomes more and more difficult to make progress. Plus, you're limited to selecting for traits that the species already has. I can't select for cattle with wool coats because there are no cattle that produce wool. Similarly, if you can't find a strain of wheat or any of its closely related cousins with resistance to a particular disease, then you can't select for resistance to that disease. Thus scientists started exploring ways to introduce new genes and traits into a gene pool.

POSSIBILITIES

The first breeding technique developed to introduce new traits to a species was "induced mutation breeding" (IMB). This involves exposing the

75

seeds and seed embryos to something that causes the DNA to mutate — a mutagen. The first mutagen was gamma radiation. Today there are a number of chemical mutagens that are used.

IMB was heralded as a limitless frontier that allowed scientists to greatly accelerate the primary vector of the evolution of species — spontaneous mutation followed by natural selection. The future characteristics of crops would be limited only by the imaginations of scientists. It's been almost a century since IMB techniques were developed, however, and the plant breeding laboratories are starkly devoid of any wondrous results. We have managed to marginally push the edges of the range of adaptation for some crops (for example, scientists used IMB to create varieties of rice that mature in areas with shorter growing seasons), and we've successfully created disease resistance, higher yields, stronger stems, and removed awns. What we haven't done is modified any crop so much that it isn't easily identifiable as that crop.[32]

It's likely you have never heard of IMB. The three-letter acronym you'll be familiar with is GMO — genetically modified organism. What IMB taught us is that the more you mutate something, the more likely the mutations are to be lethal. Each organism is a finely tuned "whole" with some configurable parameters, but the whole crashes if you try to modify it too much. The early GMO successes are very similar. Bt GMOs, for example, only include a short snippet of DNA to code for the synthesis of a specific protein that is lethal to specific members of the Lepidoptera family of insects. We have a lot to learn about how the "whole" of a plant works before we will be able to make more substantial changes, if ever. To draw an industrial comparison, we are now able to put a hood ornament on a car. But only because we found a hood ornament on a truck that we liked and were able to copy it.

Even if genetic modification never progresses beyond the ability to manipulate organisms to produce a specific molecule, there is an exciting frontier that has opened up to us. There is no comparison between the complexities of molecules that plants are able to synthesize versus what scientists are able to create in a chemistry laboratory. With genetic manipulation, we have the ability to cost effectively produce complex molecules that will be useful in many disciplines of science, including medicine. The Roundup Ready gene causes extra production

32. If you want to have a look at a database of crop varieties created by IMB, the International Atomic Energy Association maintains a database: http://mvgs.iaea.org/AboutMutantVarities.aspx.

of an important protein that glyphosate disrupts to kill a plant. The Flavr Savr tomato was created by extracting a specific DNA sequence and inserting it backwards to delay softening during the ripening process. The EnviroPig had a gene inserted to increase production of a key enzyme in the digestion of phosphorus. An e. coli bacteria had the DNA code for human insulin inserted. These are all incremental changes to existing organisms using DNA pieces from another organism.

PART OF THE SOLUTION OR PROBLEM?

The opposition to GMOs originates with a diverse set of reasons. Some people oppose directly messing with the DNA of an organism. Some opponents don't want to be exposed to the new risks created by GMOs. Some opponents oppose multinational corporations gaining increased control over our food supply through the patents obtained on GMO crops. I personally believe that GMOs are an over-hyped, over-demonized technology that is diverting research effort and money away from the true problems we face in trying to feed ourselves. Very few solutions to the problems we face in agriculture can be found in modifying plants to produce different molecules or even in modifying the plants themselves.

I'm sure you've heard the old saying "To a man with a hammer, everything looks like a nail." GMOs are the present-day hammer in agriculture. Desertification is a problem: We need to engineer more drought-tolerant plants. Saline soils are a problem: We need to engineer salt-tolerant plants. We're having a challenge killing a particular weed in a particular crop: We need to engineer that crop to be tolerant of a herbicide that kills the weed. Perhaps we should really be looking at the root causes instead of the symptoms. The desertification that claims two million acres a year isn't the result of problems with the plants that grow there; it is the failure of the entire hydrological system. The water cycle has failed, generally because of the mismanagement of the landscape by humans. We don't need to invent a plant to grow on the toxic wasteland where the fourth-largest lake by area, the Aral Sea, once provided a thriving fishing industry. We need to restore the ecosystem and water flows that filled the lake. There is no magic modified plant that is going to stop the soil degradation and erosion that threatens the ability of every continent to feed its population. Thus, we have a tool that is being widely used but will be largely ineffective in solving the root causes of our problems.

Adding to this issue, we allow genetically engineered life forms to be patented. This means that if I develop a certain plant trait by genetic engineering (GE), my ability to profit from it is greatly increased compared to the same trait if I had developed it by traditional breeding methods. Creating Roundup Ready was much more profitable for the breeders than if they had focused the same effort on breeding a higher-yielding variety or a disease-resistant trait that wasn't patentable. The same is true for research into production systems — the results generally can't be patented or monetized. This inevitably leads to a high percentage of agricultural research focused on creating GMOs. This is true in the public sector as well as in the private sector. Globally, university agricultural research programs are facing reductions in public research dollars combined with expectations of a return on the investment for the public funding they do receive.

RISKY BUSINESS

Are the people concerned about the risks of GMOs justified? Partially, but the risks are blown completely out of proportion and opponents have greatly understated the ability of the existing regulatory system to protect society. Yes, GMOs add risk to the food system; but every time we create a new variety, there is risk regardless of whether it was bred traditionally or through IMB or GE. I can breed a poisonous potato (or anything on the list of plants with toxins) using conventional breeding techniques. Induced mutation breeding creates more unknowns than GE techniques do. But we have a regulatory system designed to safeguard society from IMBs and traditionally bred varieties with new traits. It is a regulatory system built upon knowledge that has been acquired over decades. It contains tests that cover all the major dimensions of risks in foodstuffs. Are there any new proteins? Are there any allergens or allergen-like molecules? Are there any odd carbohydrates (think cis versus trans fats)? Are there significant differences in protein levels, nutrients, etc.? This is the "substantial equivalence" test that has been so disparaged. Proving substantial equivalence is not a free pass; there is substantial science behind it.

With that said, science doesn't know what it doesn't know. We now know that CFCs deplete the ozone layer, but we didn't at the time they were approved as a replacement for ammonia in refrigeration systems. We didn't have a word for bio-accumulation when DDT was approved. Now, all pesticide approvals include an assessment of the potential for bio-accumulation. The problem with the substantial equivalence testing protocols is that they are designed to detect problems that we know could occur. They are not designed to detect new problems. The analogy I draw is to the early days of anti-virus software. The software could detect any known virus but was useless against a virus that hadn't been identified and included in its database yet.

Today's computer security systems look for the behaviours of viruses as well as the specific viruses themselves. Long-term animal feeding studies are the best way to discover whether there is something happening with a GMO that we don't know about. They aren't required as part of the regulatory review of GMOs, but many scientists have completed them to provide input to the public debate around the safety of GMOs. Snell et al wrote one of the latest comprehensive reviews of long-term and multi-generational studies conducted on a number of different GMOs. The 2012 study states:

Results from all the 24 studies do not suggest any health hazards and, in general, there were no statistically significant differences within parameters observed. However, some small differences were observed, though these fell within the normal variation range of the considered parameter and thus had no biological or toxicological significance.[33]

Therefore, I don't get too excited about eating food containing GMOs. However, I'm concerned by the vitriol that permeates the scientific investigation of the safety of GMOs. Both sides seem rather unprofessional in their attempts to rebut or debunk the studies of scientific colleagues at times. The scientific method is messy. Data is interpreted, conclusions are drawn, and then other scientists retest them. The original hypotheses get adjusted, and we gradually accumulate knowledge. Unfortunately, the GMO discussion has become so politicized that true scientific discovery is being impeded due to the fear that any piece of research that questions the safety of GMOs will be seized upon by the anti-

---

33. Chelsea Snell et al., "Health impact of GM plant diets."

GMO activists and used out of context. The physicists who shook the very foundations of physics by announcing they had measured a particle travelling faster than the speed of light haven't been subjected to near the same level of public reproach — despite announcing several months later that there had been a measurement error and the particle had not in fact exceeded the speed of light. Diverse views are the strength of the scientific method, not a weakness.

MODIFICATIONS ARE FOREVER

Another aspect of GMOs that worries me is impossible to test for. GMOs are self-replicating technology that, once released into the environment, may be impossible to retrieve. When we release a GMO, we are releasing it into the "production system" to use a computer analogy. Even computer scientists who have built and fully understand how their entire system works make a distinction between the development system and the production system. The development system is a replica of the production system. All new modifications have to be proven to work in the development system before being pushed into the production system. Even if the code appears stable, there is a sequence of testing scripts that are used to test as many potential scenarios as possible. And even with these precautions, pushing a new "release" of code has been known to crash the production environment.

Biotech geneticists will argue that their modifications are contained in a laboratory until they have done sufficient testing to ensure that nothing radical will happen when pushed into the "production system" of the Earth's environment. (Remember that the ecosystem of the cell is part of the ecosystem of the organism, a part of the ecosystem of a biome, which is part of the ecosystem of the planet.) It's impossible to maintain a "development" version of the Earth to test all modifications before we push them into the "production" version of the Earth. If we accidentally crash the system, we can't "roll back" to the previous stable release while we figure out what went wrong. We have to get it right every time.

Herein lies the conundrum of genetically modifying organisms. We can see the unlimited potential for these techniques (although our expectations might be a little overblown). We have the hubris to believe we can achieve the good while avoiding the bad, and we have the greed to race forward to capture the profits from genetic manipulations. However, this technology is unlike anything we have ever played with before. When we discovered that we had a hole in the ozone layer, we stopped producing CFCs and the atmosphere healed itself. When we discovered the impacts of some pesticides, we stopped producing them and the environment has mostly healed itself[34]. When we discovered that trans fats were negatively impacting our health, we stopped producing them. When we discover a GMO is having an unintended impact on the environment, how do we pull it back?

There have already been at least three escapes of GM crops from research facilities, two of them after it was believed that all seed had been destroyed. Three unapproved GMOs have contaminated commercial grain supplies: CDC Triffid flax, Liberty Link rice, and Roundup Ready wheat. In none of the cases have authorities been able to identify exactly how the GM varieties escaped. In the first two cases, the GM varieties were found to contaminate the pedigreed seed being produced for sale to farmers, and the contamination caused the loss of export markets for the crops. In the case of CDC Triffid flax, Canadian farmers lost access to European markets. In the case of Liberty Link rice, American rice farmers lost access to Asian and European markets. The Liberty Link rice had never been approved for commercial production in the United States. Researchers and industry believed that all seed from the variety had been destroyed, but a French company accidently discovered the contamination in shipments of rice from the US. CDC Triffid flax had been approved for production of linseed oil for commercial uses but not flax for human consumption. It was detected in a shipment of flax bound for human consumption. Investigators are still trying to figure out how Roundup Ready wheat was growing in a farmer's field several years after all known seed was destroyed.

What will happen when we introduce a self-replicating life form that has unanticipated consequences into the environment? Google "invasive species." We know how this movie ends, but here's a real-life example: Kudzu vines are choking North American ecosystems. Kudzu is an Asian legume that was introduced into the US as a fast-growing plant for cattle pastures. Everything was fine while the kudzu was contained in pastures and regularly grazed, but once it escaped into the wild, it became an uncontrollable nightmare. Another example is Zebra Mussels, which have, along with excessive nutrient loading, sucked the life out of the Great

---

34. There are still farms in the US that have soil which tests above the EPA limit for DDT despite none having been applied for over 40 years.

Lakes. In still another example, the US is working swiftly to create a permanent barrier on the Illinois River to prevent the Asian Carp that have polluted the Mississippi River Basin from reaching the Great Lakes and mopping up whatever is left. Asian carp were deliberately introduced into irrigation ponds to consume the vegetation. Irrigation ponds were supposed to be foolproof containments, but when hurricanes hit the region, thousands of square miles were flooded and the carp swam wherever they chose.

Then there was the escape of Africanized bees from a research facility in South America. One worker made one mistake and one queen escaped. Africanized bees are gradually invading every ecosystem in North America and displacing wild pollinators. The escape of Africanized bees is a direct analogy to GMOs. They are a self-replicating technology that has had disastrous and ever-spreading impacts. Every precaution had been taken to prevent the escape of the queens. The researchers understood the potential consequences of their research, but they believed that it could be controlled by controlling the queens. It only took one mistake and Pandora's box was opened.

With GMOs, we have the potential to create a species that is disastrous in all ecosystems. With the rapid adoption curves of GMO traits in agricultural crops, a new trait can be planted on every continent before it has been in general release in any ecosystem for more than a year or two. By the time we have enough scientific evidence to "prove" that a new GMO trait is the source of unintended damage, it is likely that we wouldn't be able to put the genie back in the bottle. Doubt my reasoning? A subsequent chapter discusses antibiotic-resistant bacteria. The resistance was a foreseen consequence. We knew it was coming, and we still screwed it up.

How high is the risk of humanity creating a GMO with unintended consequences? From where I'm sitting, it is a 100% certainty. There is no technology that we have developed in the history of man that hasn't had unintended consequences. Just the sheer number of genetic manipulations that are occurring every day points to someone, somewhere, missing something. Avoiding calamity 99.9999% of the time isn't good enough.

TERMINATOR

People love to rally behind a cause and be against something. Very few people take the time to think through the long-term implications of stopping whatever it is they are against. Where I live, we have a government in power that has committed to shutting down all of our coal-fired generating stations. We have people vehemently opposed to bringing a refurbished nuclear reactor back online. A group was successful in stopping the construction of a new natural-gas-fired generating station. We have people opposing wind turbines. Finally, we have people opposing solar panel farms. I guess they're all planning on spending their evenings pedalling a stationary bike hooked to a generator. It's easy to be against things, to print up flyers, to create a sly acronym, to set up a Facebook page, and launch a website. However, every time you say no to something, you are implicitly choosing to do something else. If all the groups with bad cases of nimbyism with respect to energy generation here are successful in stopping the developments, they will have, in effect, "chosen" either higher electricity rates here (with higher pollution levels in another jurisdiction because we will be forced to import electricity), or rolling blackouts during the hottest summer afternoons. Will that view without a wind turbine be keeping them cool then?

This is where I part ways with a lot of people on both the organic and conventional sides of the GMO discussion. I favour "terminator technology" because it is another line of defence against the unintended escape of GMOs. The "terminator gene" is used to refer to a number of technologies that render the seed of crops sterile. It was a cause designed for sensationalist headlines and thirty-second sound bites. I don't think we made the correct choice as a society. When I went back and read the arguments for and against terminator technology, the proponents' primary argument was that it was a way to contain their biotechnology, prevent its spread, and ensure that they controlled the varieties they had the rights to. The opponents' primary arguments were that it would deliver control of the world's seed resources to the large multinational corporations and would facilitate ever more genetic manipulations. Seed saving would be eliminated. These would be excellent arguments, if only these problems didn't already exist.

We now live in a world where genetic modifications have spread to the wild populations of corn and mustard. Very few non-GMO corn or soybean varieties are planted. The multinational companies have bought up most of the varieties of most cultivated plants and animals in the world. The companies control their innovations through oppressive technology use agreements (TUAs) and civil prosecution of farmers suspected of seed saving. These companies have formed a cabal that has a blacklist of growers who have violated a TUA and are therefore prohibited from buying anyone's GMO seed again, ever.

In today's world, I don't believe that any organic grower of corn or canola can honestly claim that their crop is 100% GMO-free. Corn and canola are wind pollinated and pollen can travel a long distance on the wind. For example, there is an island in the middle of Lake Ontario where a number of corn seed breeders have their foundation plots. They provide free sweet corn to all the residents of the island all summer long so that no one will grow sweet corn on the island and potentially contaminate the field corn. For a while, one of the major canola breeders was using a farmer in Northern New Brunswick to multiply its seed because he was the only canola grower in that part of the province and there would be virtually no chance of impurities creeping into their seed supply. Those are the distances needed to truly isolate organic production from conventional production to eliminate the potential for contamination with GMOs.

Let's review. Corporate control of our genetic resources: check. Wide proliferation of GMO crops: check. GMO genes in the wild populations: check. GMO genes in certified organic crops: check. Farmers unable to save their own seed: check. Why did we oppose terminator technology again? Everything that we feared would result from terminator technology has occurred anyway and we have failed to contain GMOs.

If, and this is a big "if," the terminator trait is carried on the male side (the pollen), any time the GMO crops' pollen tried to sneak away and party on the wild side, the offspring of that copulation would be sterile and the gene transmission would be stopped right there. Organic seed crops could be grown alongside a field of GMOs without fear of the seed becoming contaminated.

The alarmists raise the concern that if all crops are terminator and there is a seed crop failure, then we won't have any seed for the following crop. This scenario is within the realm of possibility, but just barely. The North American corn industry has been almost exclusively based on hybrid seed for half a century. The corn that farmers harvest is useless to plant as seed for the following year. The plants would not yield nearly as much, and the variation in maturity and plant size would be significant. In those 50 years, we have never come close to not having enough seed. There have been years when popular varieties in a particular region have been in short supply, but we've never had a catastrophic shortage of seed. Plus, some biotech companies had developed terminator technologies that were a two-step process. The seeds produced were sterile but became fertile again when treated with a specific compound. In the case of this technology, all seeds had the potential to be fertile, but only if the owners of the varieties needed them to be.

Therefore, it is possible to conceive of a parallel universe where certified organic grains are truly 100% free of GMOs and that wild populations would still be free of GMOs had terminator technology been used. Did we pick the wrong hill to die on? I don't know, and we'll never know because, like Pandora's box, there can be no taking these genes back. However, now that we have observed our worst fears coming to fruition, it is time to look at how genetic manipulations can be contained by something other than a TUA pollen can't read. The next generation of biotech plants includes plants that produce pharmaceuticals. There is a Californian company with a genetically modified rice cultivar that produces a compound that is an effective treatment for diarrhea. The company is having trouble finding a place to run plot trials because of the rice industry's fear of a repetition of the Liberty Link fiasco. Paradoxically, the escape of GMOs with virtually no benefit for society is preventing trials of GMOs with the potential for real benefits.

I'm not naïve enough to think that terminator technology would have been perfect, but it would have been an additional layer of containment for these genetic modifications. It's akin to the shear pin in most machines that cause the machine to stop turning when overloaded before any real damage is done. You hope you never break a shear pin, but when you do, you're glad that all you have to replace is a two-dollar bolt.

TO THE FUTURE AND BEYOND

Our current systems for controlling experimental GMOs are not perfect. Both the rice and flax industries suffered millions of dollars of damage from the release of innocuous GMOs. What would have happened if the varieties were a real risk to human health? This poses more questions: How do we decide which risks we are willing to take and what safeguards we need in place before we are willing to have a new life form introduced into our ecosystem? How do you balance the possibility of curing a disease against the possibility of creating a new one? You are potentially trading existing lives for future lives. Proponents of biotechnology always point to the potential for good while downplaying the potential for negative side effects. We have no certainty of a cure for any specific condition, but we have close to a certainty of negative side effects. Biotechnology can create good, but will the price be worth it? In some cases absolutely, but in others it won't.

I have to confess, my first assessment of a GMO didn't include any of the above. In 1998, when the first Bt corn variety was introduced in Canada, my only question was: "Will it make me more money?" To answer the question, we split a field and planted half using the Bt variety and half with the identical variety minus the Bt. The reduced insect damage in Bt corn resulted in more bushels of corn in my bins. Plus, I had less frustration harvesting the Bt variety. The reduction in insect damage was immediately visible when we pulled in with the combine — fewer stalks of corn had snapped, or "lodged." (Corn harvesting equipment is designed assuming that all the corn stalks are standing upright in neat rows. Lodged corn is a tangled weave of half-upright stalks.) Less lodged corn means more corn in the tank and less time in the field. It's a double win.

GMO adoption on farms has occurred at lightning speed because they provide more benefit than they cost. Are we 100% comfortable signing the technology use agreements? No. Do we enjoy the biotech companies' enforcement actions? No. However, to be sustainable, a farm has to be profitable and every farmer is always looking for the next idea that could either lower costs or increase production. Once something is approved and available for sale, farmers have to use it to remain in business. Otherwise, they will gradually be squeezed out by competitors able to pay more rent or bid more for land because of their cost advantage. GMOs are a tool for achieving increased margins.

From a consumer perspective, increasing farmers' margins probably isn't on the top of your selection criteria for food choices. To this point, GMOs provide little benefit for consumers, but consumers bear the majority of the potential risk for harm. If you're stressed about not knowing which products in the grocery store contain GMOs because there is no requirement for labelling, relax; it's easy to tell. If the processed food label doesn't say certified organic, then you can assume that the food contains GMOs. GMOs are also starting to infiltrate the produce department.[35]

---

35. The following links to a database of plants and the status of the development of GMO varieties: http://www.gmo-compass.org/eng/database/plants/.

I'm not sure if we'll ever see many GMOs with true benefits for consumers. Biotech companies like to talk about the potential to genetically engineer "more nutritious" food. The lack of nutrition in Western diets isn't due to a lack of availability of nutritious food. The existing cornucopia of fruits, vegetables, grains, and meats can provide a perfectly balanced diet. We don't need a potato with increased levels of vitamins and minerals so that fast food fries will deliver a little nutrition — we need to eat more foods with the nutrition already built in.

One of the first GMOs with a "benefit" for consumers is an apple that doesn't brown when sliced. Moms everywhere will be able to pack apple slices for their children's snacks without fear of them turning brown. (I didn't realize this was such a widespread problem. I had an apple in my school lunch just about every day. It never browned. My mom used a high-tech, sophisticated method of keeping the apple from browning — she left the skin intact.) The reality is the benefit is really for the food processing industry — they will now be able to create snacks with the halo effect of containing "real, fresh apple pieces." I expect the vast majority of GMO food crops will be for the benefit of the food industry, just as the majority of the focus of IMB and conventional breeding programs has been for the benefit of farmers and industry.

But before you reject every GMO as a matter of principle, make sure you understand why you object to them. I want to walk you through a discussion of an existing GMO to try and understand where your objections to GMOs originate. Researchers have developed both GMO cows and goats that produce milk that is almost identical to human milk. Most people I mention this to are revolted by the thought. However, if you had a baby that needed some form of supplementation beyond breast milk, what is the logical choice? Infant formula? Or "human" milk from GMO cows or goats?

We already consume the excretions from the mammary glands of cows and goats in various forms and feed them to our children. The infant formula would contain ingredients derived from cow's milk in most cases. If you object on the basis of it being disgusting to feed your baby something secreted by an animal, both the GMO milk and infant formula fail the test.

If you object because the GMO milk is "unnatural," read the list of ingredients on a package of infant formula. There is nothing "natural" about infant formula. It contains many ingredients of plant and animal origin that have been extracted, modified, and reconstituted as something that provides sustenance to an infant but you won't find it growing anywhere. Infant formula derived from plant sources is even further from mother's milk. Mother's milk is the secretion of a mammal. The factories that produce infant formula have a closer resemblance to a petrochemical refinery than a breast.

If you object because you are suspicious of corporate control and safety testing, you have a problem. The two GMOs I'm referring to were developed by government and university researchers. The infant formula is produced by the largest food company in the world. There is at least an equal possibility of corporate wrongdoing producing an unsafe baby formula as there is producing GMO milk. The GMO milk can easily be subjected to the same safety testing regime as the infant formula. Why is there reason to distrust one product more than the other?

I think the answer lies somewhere in the field of psychology. There is comfort in knowing that millions of babies have consumed the products of a particular company and lived. There is comfort despite the fact the same company distributed misinformation by implying the infant formula was superior to mother's milk. It is sometimes difficult to explain why we place our trust where we do. When foreign cars first entered the North American market, there was a significant bias against them even though millions of people were driving those brands in other parts of the world and the cars had some significant advantages to Detroit's products. I grew up about twenty minutes from one of Honda's major North American plants. The employee parking lot was full of Chevys, Fords, and Dodges despite their employee purchase plans.

As for my choice, I was lucky. We were able to raise five healthy children primarily on breast milk. None of them had trouble figuring out how to work the milk dispenser. My rational mind can easily make the case that the GMO milk would be superior for my child or grandchild, but I have to admit that food choices are not always rational decisions. How else do you explain Twinkies?

CHAPTER 7

DRUG ADDICTS

Nature is trying very hard to make us succeed, but nature does not depend on us. We are not the only experiment.

— *R. Buckminster Fuller*

When we first started selling poultry, our customers sometimes mistook our broiler chickens for turkeys. We raise our chickens in a moveable pen on pasture using a model inspired by Joel Salatin, and we raise them big — an average male will be about three kilograms (seven or eight pounds). They are twice the size of your typical whole bird in a grocery store. The next question will be, "Do you use hormones?" We answer, "No." Then we prepare for the next question: "How do you get them so big?" We provide them with fresh air, sunshine, and a few extra weeks of growth relative to the chickens you find in a regular store. We don't use hormones in our chicken feed and neither do conventional producers, there aren't any registered to use.

Several years ago, I was prescribed a fairly addictive medication. I was only on it for six weeks and when it became clear that it wasn't working, I was gradually transitioned to an alternative. Until my medication was changed, I didn't realize I had already developed a physical addiction to that drug. Coming down off of it gave me a glimpse into the world of addictions and a new appreciation for people that manage to beat an addiction after years of use.

Industrial agriculture has many addictions — herbicides, synthetic fertilizers, and antibiotics — that will be difficult to beat. Of these, antibiotics may be the most difficult to stop simply because our current production systems and facilities have been designed assuming that antibiotics exist.

ANTIBIOTIC USE

One of the first questions that gets asked in a conversation with someone new to the idea of organic meat production is "How do you farm without antibiotics?" Most people not familiar with modern animal agriculture assume that antibiotics are only used to cure disease, the same as they are in the human population. If human medicine is the paradigm you are trying to use to understand animal agriculture, it's a logical question.

However, the paradigm for use of antibiotics in agriculture is not the same as in human medicine. The vast majority of antibiotics used in animal agriculture are not used to cure disease. They are used at a "subtherapeutic" dose, which means a dose below what would cure a disease. Researchers learned back in the 1950s that animals fed a low level of antibiotics grow faster and more efficiently. It also facilitates higher densities of animals in barns.

The switch to organic animal production hasn't been nearly as difficult as I thought it would be. Fresh air, sunshine, and lower animal densities go a long way to preventing disease. I still have a responsibility to medicate any animal that is sick. I just can't market the meat from that animal as certified organic. I'll give you some examples of adjustments that we've made to our management that have virtually eliminated the need for antibiotics and other medication.

A majority of conventional beef and lamb producers calve and lamb in January. Winter calves are larger when taken to market in the fall and presumably receive higher prices. The same is true for the lamb producers. Lambing in January also gives the lamb producer a chance to sell into the spring lamb market. In our part of the world, lambing and calving in January has to be done inside a barn to prevent the newborn lambs and calves from freezing in the elements. You have to be very fastidious in the barn to keep common infections at bay. Many producers routinely give a shot of penicillin at birth as a prophylactic to manage infections. We switched to lambing and calving on pasture in May. The sunshine naturally sterilizes the pasture, and the warmer weather means the little ones have a much better chance of getting up and going on their own.[36] We move the animals every day so there is little opportunity for disease to build up, especially compared to a few birthing pens that are used continuously in the winter.

Our system is not perfect, and we still have to medicate the odd animal. Last year we had a cow get foot rot during an especially wet spring. The last calf we medicated was a January birth that got an infection through his navel. We nursed him back to health at the time, but six months later he dropped dead on pasture. The infection at birth had formed an abscess on his liver, which burst into his bloodstream, killing him very quickly.

A lot of the debate around antibiotics seems to focus on whether you get a dose of antibiotics each time you eat conventional meat. Industrial agriculture would be happy if that's where the debate continues to focus. It's another one of those red herrings. Testing will show that a miniscule percentage of meat has a measurable antibiotic residue in it. If this debate continues, they will be able to show study after study that concludes there is nothing to worry about.

There is a lot to worry about. Subtherapeutic doses of antibiotics lead to antibiotic resistance. How long have we known this? Alexander Fleming warned us that using a penicillin dose that was too low would create "mutant bacteria resistant" to penicillin in his Nobel Laureate address in 1945.[37] He conducted research a few years later that demonstrated his hypothesis was true.

So if the same guy that discovered the first antibiotic also demonstrated how antibiotic resistance developed, how did we end up allowing farmers to deliberately feed levels of antibiotics that we had a reasonable certainty would inevitably lead to antibiotic resistance? I don't think there is an easy answer. My personal view is that it was a combination of pharmaceutical companies that saw a huge market opportunity (it is estimated that 70 to 80% of all antibiotics consumed in North America are given to animals) and regulators that naively accepted arguments that a continuous discovery of new antibiotics would limit the impact of resistance to any particular drug or family of drugs. As well, questions were raised about whether antibiotic-resistant strains of bacteria in animals would have an impact on human health. The fallacy

36. In fact, if we don't spot a new calf within 12 hours of birth, we have trouble catching it to tag it and give it a dose of vitamin E and selenium. Soils in our part of the world are deficient in selenium, which is a critical element in proper muscle development.
37. See note 7.

of both arguments is now a matter of historical fact. The endless pipeline of new antibiotics abruptly slowed to a trickle through a combination of pharmaceutical companies changing their research focus,[38] and it becoming ever more difficult to find new antibiotics.

We now understand that bacteria are rather promiscuous little beings and freely exchange DNA with different species on a regular basis. The horizontal transfer of genes between different species of bacteria is now a well-accepted phenomenon. It's been known for over a decade that the development of resistance to a particular antibiotic anywhere in the world can contribute to the development of resistance anywhere else in the world.[39] Antibiotic resistance in bacteria unique to cattle or swine is something humans need to worry about. The genetic information that confers the resistance can be transferred horizontally to bacteria that infect humans.

So now that we're here, what should we do? I can't understand why regulators in North America aren't prohibiting subtherapeutic doses of antibiotics in animal feed. It's a no brainer. It would be met with much stamping of feet and gnashing of teeth, but animal agriculture would adapt (just as it has in the European Union). Those who would know are predicting the end of the antibiotic era in human health. We've managed to destroy the efficacy of one of the most revolutionary technologies in the history of mankind in less than a century — mostly because of greed. This may sound sensationalistic, but this is one point where I feel very strongly about the abuses of industrial agriculture. When you choose cheap industrial meat, you are potentially choosing the death of a loved one from antibiotic-resistant bacteria. Simple infections become life threatening if there is no antibiotic to treat them. Believe it. It is already happening, and the rate is growing.

Why aren't regulators moving quickly to save antibiotics for human use? That's a good question that regulators need to be asked. You can bet pharmaceutical companies would be disappointed if 70% of their market were eliminated. Every facet of animal agriculture would have a significant adjustment to make. When I was a confinement hog producer, I used less subtherapeutic antibiotics than most in the industry. We only

---

38. Their research focus switched from antibiotics that were taken for a few weeks to drugs that would be taken for a lifetime — drugs to treat diabetes, high cholesterol, heart problems, male performance issues, and anything else they could convince us is a disease for which they have a magic cure.
39. Thomas F. O'Brien, "Emergence."

used medicated feed for a few weeks after weaning. Did I give up some feed efficiency because I didn't feed antibiotics longer? Probably. Did I suffer a disaster? No. Could I have eliminated antibiotics from the feed entirely? Not without making some changes that would have made the farm less profitable. I would probably have had to increase my weaning age and lower the density of the weaned piglets, which would have translated into fewer pigs produced per sow per year. I would also have needed larger facilities to house the same number of pigs. That's only speculation, as I never tried. When you're moving as many pigs through an asset as quickly as possible, you make some compromises; although, I'm not sure I would have characterized it as a compromise back then. At this point, society needs to force the issue because antibiotic resistance is progressing at an alarming rate.

A coalition of environmental groups petitioned the FDA in 1999 to revoke the licensing of antibiotics at subtherapeutic doses in agriculture. The FDA finally responded over a decade later. Their answer was "No." Their reasoning, to paraphrase: "We're too scared of the pharmaceutical companies to do the right thing, so we're developing a set of voluntary guidelines instead." Sorry, did I just use my outside voice?

Here is a direct quote from the letter explaining past experience with deregistrations from Leslie Kux, Acting Assistant Commissioner for Policy, Department of Health and Human Services, FDA, to Andrew Maguire, Vice President, Environmental Health, Environmental Defense (received November 7, 2011):

The Agency's experience with contested, formal withdrawal proceedings is that the process can consume extensive periods of time and Agency resources. For example, the first NOOHs for withdrawal of nitrofuran approvals were issued in 1971, but the final rule withdrawing the approvals was not issued until 1991. Withdrawal of diethylstilbestrol (DES) approvals became final in 1979, seven years after issuance of an NOOH. More recently, the withdrawal of approved uses of erirofloxacin in poultry took almost five years and cost FDA approximately $3.3 million.[40]

Nitrofurans are a class of antibiotics that leave a potentially carcinogenic residue in the meat of animals. It took 20 years from the time the FDA had made a determination to withdraw nitrofuran approvals until

---

40. U.S. Department of Health and Human Services, "Petition Denial."

they finally completed the task — twenty years of adverse effects and risks while two elephants wrestled. The $3.3 million in legal and other expenses is held up as a large sum of money. I wonder how that amount compares to the cost to our health system due to antibiotic resistance. From that perspective, I imagine it would look like a fairly small number. They come out of different budgets, however, and governments are silos.

Why is the deregistration process so long and expensive? The "evidence-based science" is inconclusive on whether subtherapeutic doses of antibiotics contribute to antibiotic-resistant human pathogens. Just ask Big Pharma. They have lots of trials that show there is no need to worry. Big Pharma and Big Chem have figured out how to manipulate science to tie bureaucrats up in knots. As I mentioned in the chapter on pesticides, the burden of proof is on the wrong side of the equation. Once something has been registered, the burden of proof is on the government to prove that the substance is unsafe. If it is uncertain whether something is unsafe, that's not good enough. A tie goes to continued registration.

It's easy to construct "scientific" trials in biological systems in such a way as to guarantee that the outcome will be inconclusive because there is natural variation in every biological system. If you design the trial so that the natural variation matches or overwhelms the variation between treatments, you get a statistically insignificant result. When the burden of proof is on the government, the pharmaceutical companies are allowed to use inconclusive results as support for their defence. Sometimes they even get lucky and a trial will have a statistically significant result in their favour. If you flip a coin one thousand times in a row, how many times would you expect to get ten heads in a row? Ten tails? Nine heads? Nine tails? Once. Once. Ten times. Ten times. If you do enough trials and cherry pick the results, you can come up with conclusive evidence that the coin is weighted to either side. Repeat after me: The system is broken. That's without even getting into whether the political level interferes in things they know nothing about other than a campaign contributor is complaining about unfair treatment by the bureaucrats.

In the automobile industry, it doesn't take very many incidents of a part failing to cause a manufacturer recall. Even though 99.9% of the vehicles with a specific part have shown no problem, the manufacturers are forced to issue a recall notice and fix the potential problem. In the pharmaceutical and pesticide industry, 99.9% of something showing no adverse impact would be irrefutable evidence that there is nothing wrong with their product, even if a handful of situations showed an adverse impact. The conclusion would be drawn that there is no conclusive link between the product and the adverse effect. We require greater margins of safety for the vehicles we drive than the foods we ingest.

The organic rules are a little disingenuous. My first duty as an organic farmer is to the welfare of the animal. If an animal is sick with an illness that can be cured with antibiotics, my obligation is to treat the animal. However, the meat from that animal immediately and forever loses its certified organic status. Organic recognizes that there is a legitimate role for antibiotics in animal agriculture, but refuses to include treated animals as organic. The organic system is not a closed system. We can never achieve a goal where 100% of the meat produced for human consumption is certified organic. I understand why the rules were set this way. If farmers were allowed to earn a premium on the meat from animals that had been medicated, there would have been no real economic incentive to focus on using preventative measures to reduce the amount of sickness in a herd. In my view, there is a legitimate need to use antibiotics in animal agriculture, but I would not class the majority of current antibiotic use in industrial agriculture as legitimate.

The USDA has already run the white flag up the pole without curbing the subtherapeutic use of antibiotics. Canada has trouble banning anything that isn't also banned in the US. If you live in North America, the only way subtherapeutic doses of antibiotics are going to be stopped is through consumer choice. Consumer pressure removed GMO potatoes from most fast food restaurants. McDonald's dumped one of their major egg suppliers because of public outcry over evidence of animal abuse. You can choose to purchase meats from farms that don't use subtherapeutic doses of antibiotics.

HORMONES

I think a lot of people get more excited about hormones being used in meat production than antibiotic use. They accept that animals get sick and need to be medicated, but they recoil at the idea of hormones. As I look back at history, the rise of scandals around steroids and performance enhancements in athletes seemed to bring the issue of hormones used with animals to the forefront.

Which of the following are hormones: stanozolol, oxytocin, testosterone, cholesterol, Vitamin D, or progesterone? All of them are: The first is the anabolic steroid that Canadian sprinter Ben Johnson was caught with in his urine at the Seoul Olympics. The second is produced by all mammals to stimulate contractions during birth and milk production. The third is the hormone that triggers the secondary sexual characteristics in males (also found in females). The fourth isn't technically a hormone, but it is the precursor to several major hormones in the human body. The fifth should need no introduction. The sixth is one of the hormones that govern secondary sexual characteristics and cycles in females.

Do I worry about synthetic hormones used in meat production? As an eater, not really. There aren't any used in pork, chicken, or turkey production. Beef cattle are the only meat animals with registered hormones for growth promotion. Everything we eat has hormones of some form in it. It doesn't matter if you're a carnivore or a vegan; your diet comes with a side of hormones.

At the Stoddart Family Farm, we don't castrate our male animals (cattle, pigs, or lambs). Our "intact" males have significantly higher levels of several hormones in their meat than castrated males do, including castrated males that were given synthetic hormones. For us, the decision not to castrate stems from two factors: I don't like to perform any surgery on an animal that isn't necessary, and I hate castrating in particular. There's something about holding a pig between your legs and cutting its testicles out with a scalpel that holds no appeal for me. I'm not sure whether it's the proximity of the scalpel to my own privates or the odd pig that sends a squirt of a bodily fluid right in my face.

Because we directly market all of our animals, we don't have to conform to the industry standards. The pork industry castrates because intact males sometimes have an off flavour to the meat, known as "boar taint." It's uncommon, especially in modern, fast-growing pigs, but the only way for a packer to be sure there is no boar taint is to insist on castrated males. In the beef industry, castrating solves two problems. First, it eliminates the risk of unplanned pregnancies when males and females are housed together. Second, as young bulls reach maturity, they can become aggressive towards each other, especially if you are mixing groups of cattle. The aggressiveness of the bulls isn't a problem for us because they're all raised together from birth; however, if you're trying to manage thousands from multiple sources, it's a nightmare.

The hormone used to increase milk production in dairy cows (rBST) has been controversial, if nothing else. Some countries approved its use and others did not, like Canada and the EU. The objections from regulatory agencies generally weren't related to risks to human health but rather the impact on the dairy cows themselves. Cows injected with rBST have an increased incidence of lameness, decreased body condition (unhealthy weight loss), and increased reproductive problems. For these reasons, rBST use has never risen much above 15% in the US.

Is it possible to profitably produce meat without synthetic growth hormones? Absolutely, but I also believe there bigger issues to worry about in meat production.

CHAPTER 8

ANIMAL FARM

Man is the only creature that consumes without producing. He does not give milk, he does not lay eggs, he is too weak to pull the plough, he cannot run fast enough to catch rabbits. Yet he is lord of all the animals.

— *George Orwell, Animal Farm*

"I had a whole frog, human blood, and earthworms for breakfast this morning!" Little boys everywhere enjoy trying to gross out each other and adults. Farm boys, with their advanced knowledge of bodily functions and the cycle of life, take great delight in shocking their city friends and cousins. The breakfast in question was a fried duck egg, and the list includes what a duck eats (the human blood is usually in the form of mosquitoes). Sound disgusting? You should try a swig of diesel fuel, a few huffs of natural gas, and some pesticides — also known as your average breakfast cereal.

Animal agriculture exists because animals transform inedible plants and waste into food, and animals are a great way to store protein and energy without them spoiling. That's the way it started at least. Before I go any further, I need to get two definitions out of the way. Ruminants are animals with four stomachs and the ability to digest the cellulose in grass, hay, and forages (plant matter other than the seeds). Cows, sheep, and goats are the most common ruminants. Monogastrics have a single stomach, do not digest green plant matter (cellulose) well, and require higher levels of protein and energy than a grass-based diet provides. To thrive, monogastrics need supplementation with easily digestible forms of energy and protein, such as grains or seeds, fruits and vegetables, or meat (including insects, arthropods, and anything else that crawls, wiggles, or moves from place to place). The most common monogastrics are pigs and chickens.

Traditionally, ruminants and monogastrics had very different roles in agriculture. Ruminants harvested solar energy captured in the natural vegetation and turned it into fat and protein for humans to consume as meat or milk. Monogastrics, on the other hand, were primarily fed on the waste from human consumption. When wheat was cleaned before being milled, the "shorts" were fed to chickens or pigs. Table scraps went in the hog slop bucket. Rotting or surplus fruit and vegetables were fed to the chickens and pigs.

As the agriculture and food system became more industrialized, more "by-products" were available for chickens and pigs to eat and convert into human food. The first industrial by-product was probably brewers' grain — the "waste" from brewing beer. Brewers' grains are an excellent source of protein, are very palatable, and easily digested. As time progressed, other by-products were added to the list, including oilseed cake, blood meal, feather meal, distillers' grains, soybean meal, cottonseed meal, and beet pulp. Industrial by-products could be recycled through chickens and pigs to produce bacon and eggs and fertilizer.

By the way, both chickens and pigs are omnivores, not herbivores (vegetarians). All of those chicken packages that proclaim "vegetarian diet" are simply marketing bunk trying to extract a premium from misinformed consumers who have never seen the kerfuffle that ensues after several hens spot the same grasshopper. It's as if a room full of eight-year-old boys has just witnessed a piñata full of candy burst. The chaos is instantaneous, vicious, and unrelenting until every scrap of tasty treat has been devoured.

Unfortunately, somewhere during the last half of the twentieth century, the environmental positives of pork and feather farming were overwhelmed by the negatives. By concentrating too many animals in one place, the fertilizer became a pollutant. At times, the demand for soybean meal for animal diets outstrips demand for soybean oil, and the meal goes from being a by-product to the main product of the crushing industry. In many people's minds, meat became an environmental luxury that the planet couldn't afford.

Simultaneously, ruminant livestock farming was transformed by the availability of cost-effective grain and by-product feedstuffs. Beef cattle were penned in large "feedlots" to efficiently put the last pounds and fat on (finish). Dairy cattle went from grazing on pastures to their own feedlots. Ruminants were transformed from harvesting environmentally benign perennial pastures to consuming the products and by-products of monoculture annual agriculture.

There is significant merit in the claims of activists attacking North American animal agriculture. Our current practises are a material source of greenhouse gas emissions. There are animal welfare concerns that need to be addressed. We are abusing antibiotics and assisting with the creation of antibiotic-resistant pathogens. Many workers on industrial animal farms work in dehumanizing conditions. I confess that I have committed each of those sins. At the Stoddart Family Farm, our animal husbandry practises have changed dramatically in the last decade. Our cattle and sheep are only fed grass and forages, and we work to maximize the amount they harvest directly by grazing. Our pigs, chickens, and ducks all live a pasture lifestyle as appropriate. (The ducks and pigs will explore snowdrifts if the sun is shining, but chickens are a little pickier about the outdoor conditions they will brave).

THE VEGETARIAN'S DILEMMA

One of Silvia's best friends is a vegetarian. They live close by, and our children are friends. (They were the family that looked after Cauliflower while we were away.) It is said that politics make strange bedfellows. Try explaining a beef farmer dining at the table of a vegetarian or a vegetarian dining at the table of a beef farmer. Try explaining to a seven-year-old girl why vegetarians try to create dishes that look and taste like meat but won't eat meat. Her logic went something like this: Why would you choose to eat something that was ground up, mixed with artificial flavour chemicals, processed in a factory, and then extruded in the shape of bacon when you could simply eat bacon?[41]

Vegetarians have a problem. Maybe I should call it a dilemma. A sustainable future with sustainable food production requires animals. If we were to remove animals from agriculture completely, we would destroy the planet even faster than we are currently. The only major problem facing agriculture that is diminished by a vegetarian diet is antibiotic use. Erosion would increase, water use would increase, fertilizer use would increase, pesticide use would increase, food waste

---

41. I have a picture of a box of meatless chicken wings. One of my more sarcastic Twitter followers replied to my tweet with "Better known as chicken BONES?"

would increase, carbon emissions would increase, desertification would increase, and biodiversity would be diminished.

I realize I just walked a long way out on a very high limb with those statements. Let me explain before you start sawing too hard. I'm not saying that current animal agriculture practices are sustainable. Many vegetarians, however, argue from the side of all the food that is wasted by feeding it to animals and all the greenhouse gases that cows burp out every day. That is only one side of the equation.

Vegetables have a dark side too. For the most part, they are finicky and have to be coddled. The soil has to be tilled to death to create a seedbed to grow vegetables in. Their constant need of water means very few vegetables are strictly rain fed, and so irrigation is a must. The highest rates of fertilizer application, by far, are made to vegetable fields. Vegetable fields get more pesticide applications than any field growing animal feed.

As I discussed earlier in the book, sustainability isn't about the product, it is about the system that produces the product. If we look to the natural world for a model of a system that can produce food using only solar energy and rain, we see many diverse ecosystems. However, they all have one thing in common: herbivores. Plant-eating animals perform vital functions in every ecosystem: First, they distribute and speed the recycling of nutrients. Second, they are integral to the proper functioning of the carbon cycle, especially in brittle environments. And last, they can be completely powered by solar energy, and that solar energy can come from plant material that humans can't eat.

Animals are key to creating a sustainable food system. In the two-thirds of the world that is classed toward the brittle end of the environmental scale, the vegetation that is required to ensure the hydrological cycle continues to function can be digested by ruminant animals but not humans. We may smoke some grass, but we generally don't eat it. A properly managed pastoral system enhances the environment and requires no fossil fuel inputs to convert solar energy into human-edible calories and protein. The diversity of species of plants and animals in a pasture is unequalled in any crop production system.

Pastures require no tillage and therefore the soils under them can be a carbon sink instead of a carbon source. Well-managed, rotationally grazed pastures can capture and store carbon every year. There is no limit to how much carbon can be stored under grasslands. Simultaneously, pasture systems build soil whereas vegetable and bean production destroys soil. The tillage required for vegetable production releases carbon from the soil, disrupts the soil ecosystem, and causes erosion.

I don't care whether you are talking about conventional or organic vegetable and bean production. Organic substitutes more tillage for the herbicides in conventional agriculture. The soil-building cover crops used in organic production are invariably tilled into the soil. Those same cover crops could be grazed by ruminants before being turned into the soil. The compost that is used for fertility in organic production is a waste of animal feed and energy. Anything compostable can go through an animal, produce meat, and be turned into plant-available nutrients much more efficiently than composting. Composting requires energy to maintain the proper aerobic conditions (when done on a large scale, the energy is fossil fuels powering large machines), and nutrients are lost to the atmosphere during the composting process. When in a sarcastic mood, I've been known to define composting as mixing animal feed with diesel fuel to create plant food. I'd rather have bacon than a diesel fuel bill.

To get from the unsustainable animal production systems that we currently have to the systems I am describing requires a lot of change. I'm not describing tweaks to the current industrial meat production systems; I'm describing a complete overhaul. The new system does not have huge feedlots of cattle, trainloads of corn are not sent to huge confinement animal feeding operations, and self-harvest is maximized, as is feeding of waste. Manure switches from being a liability to an asset. It is a system that is distributed and landscape appropriate. It is a system that optimizes the flow of energy and the cycling of nutrients rather than the yield at any specific point.

These changes will simultaneously improve how animals are treated. I hope that those vegetarians who object to eating meat because of the treatment of animals see in the system I am describing a dramatic change in the welfare of the animals.

ANIMAL HUSBANDRY AND WELFARE

Husbandry is a quaint term that is rarely used anymore. The Department of Animal Husbandry at my alma mater, the University of Guelph, had its name changed to the Department of Animal Science long before I attended. The only evidence on campus of the word is imprinted in the masonry of Zavitz Hall — the former Field Husbandry building — now housing the Department of Fine Art and Music.

If you look up the definition of husbandry in the context of animals or crops, you will find a definition similar to these:

1.a. The act or practice of cultivating crops and breeding and raising livestock; agriculture.

b. The application of scientific principles to agriculture, especially to animal breeding.

2. Careful management or conservation of resources; economy.[42]

or

1. (Life Sciences & Allied Applications / Agriculture) farming, esp when regarded as a science, skill, or art

2. management of affairs and resources[43]

There's a lot rolled into the word — lot more than the term "animal science" conveys. The application of science to agriculture is imbedded in husbandry, but it doesn't stop there. There is an implied duty of care for the animals, crops, and resources. The second definition makes reference to "skill, or art." There is definitely something beyond science in successfully farming. Science likes repeatability and precision, but agriculture involves biological life forms that do not always conform. A farm isn't a laboratory where repeatable experiments can be conducted. Growing conditions are rarely perfect. You are always juggling competing priorities with less than perfect information. As much as we try to break agriculture down to randomized block design plot research, not everything scales up to the real world.

I believe animal husbandry starts with an understanding of an animal's perspective on the world. All domestic livestock were domesticated from prey animals, and they have eyes set on the sides of their head, giving them excellent peripheral vision but very poor depth perception. I know I've raised the concept of perspective in relation to many concepts in this book, but I can't stress how many times I have realized that something or someone I was cursing as "stupid" was reacting in a perfectly logical manner from their/its perspective.

42. *The American Heritage Dictionary of the English Language*, 4th ed., s.v. "husbandry."
43. Collins English Dictionary: Complete and Unabridged, s.v. "husbandry."

When we had the confinement hog operation, we were a "continuous flow" operation, which meant we marketed pigs every week. We usually loaded on Sunday afternoon or very early Monday morning because we were close to the packing plant and our pigs were needed to start the Monday morning shift. When we first started loading out of the renovated barns, we fought and manhandled just about every pig that had to be loaded up onto the stock truck. Over time we learned that if we had the pigs out of their pens for 10 or 15 minutes before the truck arrived and the pigs had a chance to explore the aisle and the ramp, they would load with a lot less effort. (We still always had an electric prod in our back pocket to "encourage" a reluctant pig to move forward.) We also learned that moving pigs down an aisle with solid sides was easier than an aisle with "spindle" gates, which let the pigs see through into the pens on either side.

There was one particularly perplexing problem that took us quite a while to understand and overcome. Our pig-finishing barn was an old-style barn that used straw for bedding and had a gutter cleaner that went through the back of each pen. The gutter cleaner was essentially a strong chain with paddles attached in a continuous loop, which pushed whatever was scraped into the eight-inch-deep gutter at the back of the pens out of the barn and dropped it onto a pile in the barnyard. When we first renovated the barn in 1979, we had a problem with pigs getting into the gutter and using their noses to flip the paddles up. We'd have to go around and make sure all the paddles were horizontal before starting the gutter cleaner. Sometimes, if you were working alone, you would put all the paddles down and run to the "on" switch, but before you got there pigs would have already flipped some paddles up. Dad's solution was to install an electric fence over the gutter. It was rigged up identical to the electric fences that we now use on pasture. It was very effective at keeping the pigs out of the gutter and lowering our repair bills and frustration level.

The electric fence caused another problem. There was one place where the loading aisle crossed the gutter cleaner. We had a plank "bridge" over the gutter, which fit quite snugly and was virtually indistinguishable from the cement floor on either side. The pigs refused to cross it. At first we thought the pigs could sense the difference between the soundness of the bridge versus the cement floor of the barn. It was only 18-inches wide, but pigs would stack themselves two deep to avoid crossing it. Using the electric prod often made things worse. We finally realized that because pigs have eyes on the sides of their heads they were getting a full view of the electric fence through the spindle gates. Despite the fact that there was no fence across the aisle where the pigs were, they were seeing the fence in their peripheral vision. We put up a solid panel along the aisle to block the pigs' view of the electric fence and our problems disappeared. It seemed so obvious once we figured it out, but since we weren't looking at the situation from the perspective of the pigs, we missed it.

Now, we load pigs without breaking a sweat. We've switched from a stock truck that was almost four feet off the ground to a stock trailer that has an 18-inch step up. We call the pigs every morning when we feed them and they come running. They will follow someone with a white bucket of feed to hell and back. We make sure they're hungry when we want to load them and we can swiftly have them walk right into the trailer. Granted, we're not shipping thousands of pigs. I don't even own an animal prod. Even producers that have them try to avoid using them because they can cause a bruise at the point of application, generally in the ham of the pig.

Cattle are another challenge. Most farmers that handle any quantity of cattle have a handling facility that is designed to contain the cattle and load them with the least stress possible. A good facility starts with a spacious crowding tub so that cattle will walk in without balking. The crowding tub is generally a semicircle with a gate that swings behind the cattle and encourages them to move forward and into the chute to be loaded. This type of facility means cattle can be handled safely and without an excess of stress on the animals.

While we have a proper cattle handling facility, we've increased our skills working with cattle to a point where we can now sort and load cattle straight off a pasture if needed. We manipulate cattle using a combination of flight response and their respect for electric fences. To get cattle to move straight ahead, you have to pressure them from the side so they can see you in their peripheral vision. If you pressure cattle from directly behind, they will constantly turn so that they can see you. Because we rotationally graze, our cattle are in a different spot everyday. Many of the pastures are a distance from the barns and the handling facility. To load out of a pasture, we pull the trailer right into the pasture and then set up a mini-corral with a high-visibility electric fence (with no charge in it). We create a funnel to channel the steers we want into the corral. We then divide the corral in half and sort the animals we don't want out. Once we're down to the animals we want, we slide the trailer door open and gradually collapse the corral down so the cattle see walking on to the trailer as their only option. They walk on, and we slide the door shut behind them. This approach only works if you and the cattle stay calm. If cattle are excited, they will run straight through a single wire fence. We have an advantage with our rotationally grazed herd. They see us every day and they see new fences every day, so they're not upset by our presence or confused by the sudden appearance of a fence in a place they've never seen one before.

This is a sharp contrast to what happened one November Sunday in my youth. My parents had cattle on a rented farm two and a half kilometres down the road. It had snowed heavily and we needed to bring the cattle back to the home farm to sort which ones were being sold and which ones we would overwinter on the home place. We were renting the farm from an individual who had no experience with farming or livestock but fancied himself a farmer because he owned a farm. This individual thought he would "help" us by rounding the cattle up into the barn corral before we got there. He decided the best method to round up the cattle was to use a dirt bike. The cattle had never encountered a dirt bike before, and all he managed to do was completely spook them so that they didn't even respond to us properly. We did finally get the majority of them into the corral and loaded, but several bolted. We were wading through deep drifts of snow, trying to follow them as they went gallivanting across neighbours' farms. We caught one when it attempted to walk across the tarp over a neighbour's winterized inground pool. The last one was finally captured when it was convinced to go into a neighbour's barn five kilometres from where we started. The moral of this story is that cattle, or any animal for that matter, can become very excited and unpredictable when frightened or handled by an amateur.

I've had our entire herd of 75 head get out because a deer knocked some fence down. I simply enclosed the cattle within an electric fence perimeter and then called them back to where they belonged. They calmly followed me. I didn't even have to call the house for help. On our farm, the cows come when I call them and the sheep come when Silvia calls them. Neither of us is nearly as effective calling the other. The cows know me because I generally move them from one pasture to the next, and I make a point of calling them every time I move them so they associate my call with something desirable as opposed to them only hearing my voice when I'm cursing and angered. Similarly, Silvia is the primary person that cares for our flock of sheep. There are a couple of ewes that always come to see if Silvia has some grain or another treat. As soon as one ewe is following and calling for feed, the rest of the flock soon follows.

I wish all people involved in livestock handling were as skilled. I've been at the stockyards and visited the processing plants when large numbers of animals have been moving through. It's easy to spot the people with experience and confidence moving animals. They're calmer, and the animals are calmer. The guys who are afraid of the animals stir the animals up with their actions. It is absolutely true that animals can sense your fear, anger, or other strong emotions. There have been a few occasions when I've been short of time and lacked the patience to move or sort the animals calmly and at their pace. They immediately sensed that something was "up" and balked at doing what I wanted. A couple of times I've had to walk away and come back after I'd calmed down. I confess that a couple of times my family has walked away from me and come back once I'd calmed down. I'm not perfect, nobody is.

How does all of this impact you as an eater? The first impact is simply the quality of life for the animals. It's easy to tell how animals have been handled simply by observing them at a farm. When we bought our herd of White Parks, it was clear when we first went to the farm to see them that they hadn't been treated well. They were skittish, making it difficult to walk among them. Once we bought them and brought them to our farm, it was over a month before I could walk among them. Initially they would run to the opposite side of the pasture. I bribed them with a small amount grain. At first they would wait until I left before having a snack, but eventually some of the braver ones would come up as soon as they saw me. Finally, they would allow me to walk among the herd. There are a few that are still uneasy. Molly has a particular distrust of humans. As soon as anyone approaches the herd, she calls her calf to her and then stands between her calf and the person, and stares. We call it the "death stare." We know it is a death stare because she has made her intentions very clear when we've gotten too close to her calf.

Modern-day industrial agriculture likes to argue that if animals don't know anything other than the conditions they are living in, they can't be stressed because they don't know any better. I disagree. There are basic components of animals' behaviour that are inherent regardless of how they are raised. For example, pigs are inherently curious animals that need activities to occupy themselves. In the natural world, they root through the earth and chew on whatever they find. When you coop them up in modern confinement barns in cement pens with nothing to root or chew on, they develop bad behaviours. Most hog operations "dock" (cut off the end of) the tails of pigs shortly after birth. The reason? The very tip of a pig's tail has no nerve endings. Bored pigs will start chewing on each other's tails. By the time a pig feels the other pigs chewing on its tail, it's too late. The skin has been broken and the other pigs have the taste of blood. At a minimum, some pigs in the group will end up with infections. The worst-case scenario is some pigs will end up dead. Thus, tails are docked so that a pig feels and responds to the first time another pig wonders what its tail tastes like. We don't dock tails because we don't need to. Our pigs live in an environment that provides lots of opportunities for the pigs to act out their natural "pigness."

Many dairy farms today house their cattle 24/7 in what are called "free-stall" barns. The cows are free to walk around and choose which stall they lay down in. However, the cows never go outside to pasture. Their feed is delivered to them in the barn, and they walk to the milking parlour to be milked two or three times a day without ever going outside. Do those cows miss being outside if they have never experienced the outdoors? It's an interesting question that can't be answered. The conventional thinking is that production is correlated with animal contentment. Higher production is only possible with healthy and content animals. However, while the dairy industry is achieving higher daily production from cows, total lifetime production is decreasing because cows aren't producing for as many lactations as they used to. If I follow the conventional arguments, I would take that as a sign that modern industrial methods are stressing the animals more and their productive lives are being shortened.

If you have ever been on a farm at the moment the cows are let back out onto pasture for the first time in spring, you will witness cows expressing unbridled joy. They literally go skipping across the pasture, jumping and running. It is a site to see. So then the question changes from "Do cows miss being outside if they've never been?" to "Do we have the right to diminish a cow's life experience by denying it something that it clearly enjoys when given the opportunity?" I don't think so. I don't think we have to. Confining animals is partially about efficiency, but a large component is about convenience for the industrial food system. You can't slaughter the same number of pigs every day if they aren't raised in confinement barns because there are times during the year when you don't want to have young piglets due to the weather. Chickens' egg laying cycles are influenced by day length. When you put them in a barn with total control of the lighting, you can keep their lay fairly consistent over the year. Our production always falls off in autumn, as the days get shorter even though we have supplemental lighting on timers to lessen the impact of shorter days. In the industrial system, the pipeline of product needs to flow as consistently as possible. Some egg producers are even experimenting with compressing the light/dark cycles to create an extra day each week to increase production.

It might strike some of you as strange to be discussing the quality of life of an animal that we are ultimately going to kill for food. Do we have the right to raise them for the sole purpose of killing them? Philosophically, it is an interesting bargain that a domesticated species strikes with humans. As a species, they are much better off than if they had continued to be wild animals. Humans have grown and collected feed for them and transported them to every continent except Antarctica.[44] Each individual within the species has a determinate life expectancy. The probability of the animal living to maturity is greatly increased in the case of the domesticated animal as compared to the wild animal; however, those that live to maturity may have longer lifespans in the wild.

An additional consideration is the death the animal would experience in the wild versus the methods employed by humans. In the wild, most animals die a very violent death at the teeth of their predators. As humans, we claim humanity in the way we dispatch our food. Yet there is much room for improvement. The large industrial animal processing facilities, for instance, employ assembly line techniques to disassemble the carcasses. Those lines have a fixed speed that pushes the ability of workers to feed it. There is no way to stop to correct a mistake. With our cattle, we prefer to use small abattoirs that don't have lines and have a greater opportunity to be humane.

I'm not going to go into the specifics of all the ills of confinement animal production — there are other well documented sources on that topic. I will warn you that while the activists are sensationalizing things to draw attention, they shouldn't be dismissed as crazies. There are good reasons why big agriculture is slowly convincing legislators to enact laws to protect farms from investigative intrusions. They are afraid of what you might think if you knew the truth. Yes, every major food manufacturer has a code of ethics. Yes, there have been some high profile cases where an "astonished" food company abruptly switches suppliers because of allegations brought forward about misconduct. But

---

44. All domesticated cattle trace their ancestry back to one of two domestications of wild species: Bos taurus in Turkey and Bos indicus in India.

those barns and workers didn't abruptly stop churning out product. They just changed names and started supplying someone else.

Industrial animal farming will do just enough to appear to be improving conditions for the animals. The density of laying hens in cages has recently been adjusted to improve animal welfare, but the hens are still in cages too small to fully spread their wings and their beaks are still trimmed to prevent them from harming each other. In the pork industry, I expect they will reluctantly agree to phase out gestation stalls — pens that house sows for most of their lives that are only big enough for the sow to stand up and lie down. They will continue to "dock" tails, however, and they will still be cutting the testicles out of male pigs for no good reason other than the industrial food processors don't want to take the risk that an intact male has boar taint.

If you want human- and animal-friendly agriculture then you need to accept biological variation and refuse to have a factory machine do all the preparation and cooking of your meals. To provide standardized meals that can be microwaved on "high" for ten minutes, the industrial food system requires industrial farms to churn out identical products. A car assembly line can be incredibly efficient because every car going down it is identical. An animal disassembly line gains efficiency when the animals coming down it are as close to identical as possible. You don't know what you're losing. The standardization of food is also the "blandardization" of food. We almost lost the Berkshire breed of pigs because it has black hairs that slow processor line speeds. Berkshires, however, handily won a taste test conducted by the Ontario provincial pork producers' association. The industrial food system still isn't interested in Berkshires. Bland meat is more profitable — they can also sell you a magic packet of spices to add flavour to the meat.

If you're relying on a label on your food to tell you how it was raised, you're completely missing the point. A chicken farmer neighbour had a private brand line of broiler chickens available in local supermarkets with a label that said, "Free run." I don't know how many consumers were sucked in by the clever marketing and thought they were buying something raised differently to standard broiler chickens. The label was technically correct, the chickens were free to run around the barn with their ten thousand friends just like in every other conventional chicken barn in North America. Consumers were confused between the laying cages used in chicken egg production and the broiler barns where chickens roam freely in a large group in a large barn. Even certified organic production had trouble writing rules that forced producers to adopt systems that had broiler chickens outside being raised the way consumers pictured. The first rules used the phrase "access" to the outdoors. I saw conventional chicken barns with "porches" built at the doors on each floor of a three-storey barn to convert them to organic. Regulators are working to plug holes in the regulations as fast as they can, but don't assume a label means anything unless you've been to the farm that produced the chicken.

The confinement animal industry loves throwing red herrings into the discussion:

1. Confinement barns are climate controlled and protect the animals.

2. Poultry barns prevent chickens and turkeys from getting avian influenza (bird flu).

3. Turkeys will drown if they're left out in the rain.

4. Barns protect animals from predation.

5. Gestation stalls keep sows safe from being picked on in large groups.

There's enough truth in these statements to make them believable, but producers across the globe are proving there are better models. Joel Salatin and his pasture-based systems are legendary. We've adapted his methods for chicken and egg production to our local conditions. Non-confinement systems for beef and lamb production are no-brainers. Through rotational grazing, we can raise animals in conditions that work with their natural biology within an environmentally enhancing system.

Here are my answers to the red herrings:

1. Confinement barns are climate controlled and protect the animals.

   The climate is controlled, but the air isn't nearly as fresh or as healthy as on pastures. Animals do need shelter from extreme weather, but most days our animals choose to spend a good portion of their day outside.

2. Poultry barns prevent chickens and turkeys from getting avian influenza (bird flu).

   All documented cases of bird flu in Canada have occurred in confinement poultry barns. When the province of Ontario was asked to implement regulations forcing all poultry flocks, regardless of size, to be housed in doors 24/7, the province found no documented scientific evidence to justify the regulation.

3. Turkeys will drown if they're left out in the rain.

I can't believe turkey farmers still use this line. There are wild turkeys. They don't drown in the rain. I've successfully raised many turkeys outdoors. I've lost some to interesting circumstances, but the cause of death was never drowning in the rain.

4. Barns protect animals from predation.

This is true; however, electrified fences, guardian dogs, and rich biodiversity also protect animals from predation. If you never left your house, your chance of dying in a car accident would be greatly reduced. I can't say you would be better off.

5. Gestation stalls keep sows safe from being picked on in large groups.

Pigs are very hierarchical, and timid sows can be picked on; however, solitary confinement isn't necessary. Stable groups combined with modern, computerized feeding technology and good management can allow sows to be raised in loose housing while overcoming aggression.

WORKER WELFARE

I worked in six different office towers over the course of my seven years of full-time employment off the farm. It was interesting watching the transformation in thinking about office design over that time period. The first three were built using the old standard configuration — the partners' offices were on the outside with the cubicles (cube farm) in the middle. The partners had all the windows and natural light, and the poor peons made do without a view or windows. Being a country boy, having to walk into someone's office to see what the weather was doing was pure torture. The fourth office I was stationed at was a new office in a relatively new building. The configuration was reversed. The partners had offices in the centre, and the cube farm was transformed into a spacious, naturally lit work area with the cubicles eliminated.

Today, my office overlooks our pond and about half of our farm through a big bay window. For four to five months of the year, the windows are open and I have a fresh breeze flowing through along with the sounds of nature — although today the tree frogs are singing in a way that sounds very similar to the starting noise a fax machine makes. On my favourite days, my office is out in the fresh air in the fields. A few clients have been surprised when they realize that I'm standing in a pasture while talking to them on the phone. (Ewes and cows can't be shushed quite the same as children.)

If I wound the clock back to the time before I joined the office tower club, I worked in a confinement hog barn. It was my choice, since I bought it from my parents. This might surprise you: I preferred the life in the windowless office to my life in a windowless barn. I would shower before I went to work, not after I came home and before being allowed into the living spaces of the house. My kids starting hugging me immediately when I came in the house. The persistent "snuffling" cleared up. The truth is the office tower, even with all its potential health issues, was a healthier place for me than a confinement hog barn. My mother suffers from chronic obstructive pulmonary disease (COPD) as a result of almost 30 years spent in that environment. The list of farmers suffering from emphysema is long. All this so eaters can have a cheaper pork chop.

If you're eating food produced in the good ole United States of America, there's a good chance a portion of the labour was supplied by illegal immigrants. They make great farm workers. They don't make any trouble, and they will do anything asked of them. A university classmate of mine married a dairy farmer here in Ontario. About ten years ago, they decided to sell their farm here and move to the US. They traded an average Ontario dairy farm (milking 70 cows) for a small dairy farm in the Dakotas (milking 500 cows). Homeland Security busted them a few years ago for having illegal Mexican workers. Not much came of it, and not much has changed. The state and federal governments in the US know it is happening and mostly look the other way because there aren't enough American citizens desperate enough to do most of the jobs.

I may have employed some illegal workers at one point. We had an organic field of soybeans where the weeds had grown past the point where we could remove them with a tractor and row cultivator. Someone I knew gave me the number of someone who could provide contract manual labour at a reasonable rate. A few days later, I had an able-bodied crew hand weeding that field. Only one person spoke anything resembling passable English, and I didn't get a warm, fuzzy feeling when I saw the boss come to pick them up. They finished the job in a couple of days, and I never had a need to hire them again. A few months later, a local vegetable packing operation was raided and accused of having a number of illegal workers. I'm pretty sure they had used the same labour "service."

Many of you will have heard of "interns" working on organic farms. Interns exchange their labour for an education in farming and a small stipend. Most of you wouldn't do their work for your current salary. It's part of the model that organic farming works on. If we had to pay minimum wage for all the labour, prices would have to be even higher to generate a profit.

Farm workers experience everything you worry about in your food supply much more acutely. You worry about pesticide residues in your food; they worry about inhaling pesticides directly while using them. You worry about the creation of super bugs on animal farms; they worry about being infected by those superbugs. You are disgusted by the smell as you drive by a confinement animal feeding operation; they work in those conditions every day. Your choices create those conditions whether you realize it or not. The Western lifestyle was built on slavery. Initially it was peasants, then it became slaves in far-off colonies, and now it's "paid" workers here and abroad doing jobs in conditions that very few of you would tolerate.

If your soul objects to the images of animal abuse in the hidden camera videos, think of the damage that has been done to the soul of the person committing the abuse. The majority of workers on animal farms are not twisted psychopaths that enjoy abusing animals, and the majority of them do not directly abuse animals. Most of the conditions that they work with the animals in are a function of the demands of an industrial food system and continual cost pressures.

Don't limit your criticisms to large "industrial" farms. "Small" is not, by definition, humane. I've seen animal living conditions on small idyllic farms that were worse than modern barns. Before we had our own sows, we purchased weaner pigs from smaller farms that had rare breeds. I only had to look at the weaners to see that everything was not tiddly-poo and copasetic on some of those farms.

I remember taking a tour to one of the "darlings" of the sustainable food movement. The farm had been lauded for its pasture-based systems for hog rearing, use of rare-breed hogs, and the high percentage of forages in the pigs' diets. I was sceptical based on the preceding section on monogastric diet requirements. What I found didn't surprise me. The sows were being milked down to skin and bones because there wasn't enough energy in their diets. The farmer's description of his rotation as using the pigs to "plough" the fields before the next crop was planted was a whitewash. The pigs were destroying the soil structure and leaving soils open to severe erosion. It's true that the pigs were on pasture, but I'm not sure it was benefitting them or the pasture.

Soil damage and erosion are aspects of pasture-raising hogs that I can't claim to have completely figured out yet. I'm breaking up and reseeding an eight-year-old pasture that was destroyed by the pigs last fall and this spring. Even the pasture they were on last growing season is a rough and bumpy place due to their rooting. When I consider the many dimensions of environmental impact and sustainability in hog farming, I really wonder if the best option for raising pigs is large concrete slabs with a large pile of organic waste to root through. We're working on sourcing more of our pigs' diets from waste streams. It would be easy if our customers didn't object to GMOs and I didn't object to the ethanol industry — we could have a tractor-trailer load of distillers' grains from a local ethanol plant delivered whenever we wanted. We're looking to improve our choices, but I expect it will take a lot of explaining to new customers as to why we choose to feed garbage to our pigs rather than organic grains. We'll have to get a better name for it. Joel Salatin has his "Pig-aerator Pork." How does "RepurPork" (as in waste-repurposing pork) sound?

If you are trying to choose a meat supplier for your family, you need to ask questions to get more information than the standard package label will contain. If you're purchasing beef or lamb, you want 100% grass fed. It is the healthiest for you, the animals, and the environment. There is no excuse for feeding grain to cattle or sheep, other than it makes it easier to manage and grow them. To identify a true grass-fed animal farm, ask some questions about their grazing program. A farmer supplying 100% grass-fed beef or lamb will be using some form of rotational grazing. If you're comparing farms, the number of days on pasture is a good comparator. Longer generally means better management and more carbon sequestration.

For pigs and chickens, you're looking for "pasture raised." There is no such thing as a grass-fed pig or chicken. Listen for phrases such as "chicken tractor," "egg-mobile," "hen-wagon," or "pasture pens." The farmer should be describing some system that moves hens and broiler chickens around on a pasture. Having them in a barn with a door open to the yard doesn't provide much benefit if the yard is completely denuded of vegetation. Once you've had a true pasture-raised egg during the green season, you will be able to identify them just by cracking one open — there is no other way to get that rich, almost-orange yolk.

It's fairly straightforward to connect your grocery choices to sustainable agriculture and farmers, but you also need to consider your dining choices outside the home. When you choose to eat fried chicken from a franchised fast food restaurant over cooking at home, you are voting for industrial farming that needs to turn out identical chickens everyday so the cheap labour in the "restaurant" can operate a machine that deep fries the chicken for exactly the same length of time every time and still have the chicken cooked perfectly.

Industrial agriculture is needed to feed the industrial food system. To produce identical chickens or eggs or pigs every day of the year requires removing as much biological variation out of the system as possible. To ensure a hen lays consistently throughout the year, she is put in a barn with completely artificial lighting so that changing day length won't trigger her biological clock to stop laying when the days get short. To ensure cows produce the same amount of milk consistently so the cheese factory can churn out cheese consistently, they are locked in a barn and fed an identical ration every day. If the cows harvested their own feed from a pasture, the changing nutrition of the grasses with the changing of the seasons would influence the quantity and quality of the milk they produce. A cheese factory expects every hectolitre of milk to produce exactly the same quantity and quality of cheese every time because a consumer expects each block of industrial cheese to be identical every time they purchase it.

I doubt many of you would eat the whole frog, human blood, or worms I described at the beginning of the chapter, but most of you would enjoy a duck egg (especially if I baked it into a soufflé — duck eggs make lighter soufflés than chicken eggs). The judicious and appropriate use of animals in agriculture can turn many things that none of us can eat into healthy, life-sustaining food. The challenge is getting past the industrial mindset. Even classically trained chefs are challenged by variable food. I participated in a Slow Food event that featured our various eggs. We have different breeds of ducks with distinct sizes of eggs. One of the chefs commented that she had weighed each egg

116

individually to adjust her chicken egg recipe. But weight of the eggs was only one variable; they also had varying percentages of yoke and different consistencies of albumen. I also expect the eggs would have varied over time depending on whether there was good frog hunting. Our eggs are very different from the eggs that come from genetically identical chickens fed identical rations every single day. The industrial system sacrifices exceptionality for consistency.

Choose exceptional. Choose sustainable.

CHAPTER 9

BIODIVERSITY

We should preserve every scrap of biodiversity as priceless while we learn to use it and come to understand what it means to humanity.

— E.O. Wilson

To keep every cog and wheel is the first precaution of intelligent tinkering

— Aldo Leopold

The first time we received an order of day-old chicks from rare-breed chickens, I was amazed by how different the chicks from the various breeds were — colours ranging from yellow to black in both solid colours and patterns. I had the stereotypical fluffy yellow chick imprinted in my mind. I'd been around farming for almost 40 years and had never seen anything other than fluffy yellow chicks. I had the same reaction when we received our first Khaki Campbell ducklings. They have khaki feathers, khaki bills, and khaki feet. I laughed. We'd raised Pekin, Muscovy, and Moullard ducks before, and they were all fluffy yellow ducklings.

I still get surprised by the diversity that exists on our farm, including the varieties within the domesticated species we raise. Extending beyond the farm to the surrounding ecosystems, the diversity of flora and fauna is amazing — even when only observing them on a human scale. Diversity is a concept that has extreme value regardless of the context: A diverse diet is healthier. A diverse economy is more resilient. A company with a diverse workforce is less prone to groupthink. It's unfortunate that we even needed to coin the term "biodiversity." A biological system is, by definition, diverse.

A lot of the agriculture and medical technology humanity has developed is dependent on the diversity of biology. Antibiotics work because of the differences between the biology of various organisms. Herbicides work because different plants rely on different metabolic pathways. We can use insecticides in our homes because insects have different metabolic pathways than higher-order vertebrates. We were able to take a wild beast (*Bos taurus*) native to Turkey, domesticate it, and adapt it to conditions across the world precisely because of genetic diversity. If all *Bos taurus* were identical and perfectly adapted to the environment in Turkey, we wouldn't have been able to create the Angus and Highland cattle adapted to life in Scotland, or the Limousin and Simmental cattle adapted to being beasts of burden in Europe. If all plants and animals had identical metabolisms, herbicides, insecticides, and antibiotics wouldn't be possible. Yet industrial agriculture is heading in the opposite direction of diversity because efficiency is often the enemy of diversity — efficiency that is derived from standardization.

The very foundation of our success in agriculture is the biodiversity that we are gradually destroying. We have used genetic engineering to create rice with high levels of Vitamin A — Golden Rice — to replace the Vitamin A that was lost from the diets of many people in Africa because of a loss of diversity in the plants they were eating. The indigenous diet based on indigenous crops had sufficient levels. It was only when the majority of those plants were replaced with monoculture grain crops that Vitamin A deficiency became an issue.

Both the biodiversity of the animals and crops that we grow in agriculture and the biodiversity in the natural areas that surround and intersect with our farms are key to our future. I can't tell you how the loss of any particular creature will impact us exactly, but each creature inhabits a unique niche in the wild and a unique point in the food web. I'm not sure how many threads we can lose before the garment falls apart. Genetic diversity on farms is critical to long-term sustainability and resiliency.

DIVERSITY AND BREEDING

Conventional breeding techniques select the animals with the most desirable traits from each generation to form the breeding animals for the following generations. Over a period of time, a particular trait can be significantly improved.[45] The "improvement," however, comes at the loss of genetic diversity. For example, if you select for increased size generation after generation, you will get larger animals because you will have removed the genetic variation that kept animals smaller. Yet genetic information is extremely complex, and not all desirable traits can be seen or measured. It is possible that desirable traits are highly correlated with undesirable traits; therefore, at the same time you are selecting for a larger animal, you could also be selecting for something that weakens the animal.

This has occurred in the hog industry. Breeders have been focused on producing a leaner, faster-growing, more efficient pig since shortly after World War II. In the 1980s, a problem called PSE (pale, soft, exudative) pork began showing up with increasing frequency. The meat of these animals looks lighter in colour, is not as firm in texture, and is watery. The gene was eventually identified, and a test was developed to work on removing it from the gene population. The problem had been created because the gene for PSE and the genes governing leanness are highly correlated, only you can't see PSE when selecting live animals.

Another example is found in the poultry industry. Hens have been selected for egg production for over a century, with immense genetic pressure applied in the last half of the twentieth century. These birds worked well in battery cages where there were only five or six birds, but with the animal welfare concerns being raised about hens living their lives in a cage, which is insufficient for a hen to fully stretch her wings, researchers have been studying large-group housing arrangements. They were surprised by the level of aggression between hens and the continual fighting. A large group of a thousand or more hens is no more natural than a battery cage. It also turns out that bad attitude is correlated with egg production. The term "pecking order" comes from exactly what it sounds like — chickendom is a highly structured society where every hen knows her place. However, it is one task to sort out the order in a flock of 25 or 50 birds. Sorting out pecking order in a flock of several thousand is several orders of magnitude more difficult and is in fact impossible for hens to achieve. Researchers are now developing lines of calm hens to cross with the high egg producers to reduce the stress in large-flock housing situations.

When you look around the cattle industry, it's hard to believe that all the various breeds, sizes, colours, milk, and beef traits arose from two

---

45. Traits are individual characteristics of the animals, such as size, leg strength, leg positioning, milk production, and growth rate.

wild populations that were domesticated. All of the English and Continental breeds trace their history back to *Bos taurus*, which was domesticated in present-day Turkey. All of the "tropical" breeds trace their history back to *Bos indicus*, domesticated on the Indian subcontinent. The genetic information needed to create the vast variety we have today was contained in those two wild populations. From there, farmers in various regions selected the animals that were best adapted to local conditions and purposes. Sadly, over the last 50 years, all of that diversity has been shrinking. Three global genetics companies control a large percentage of the world's meat chicken, egg chicken, and turkey genetics. Cattle, swine, and fish aren't far behind.

The dairy industry has focused on the Holstein breed. They were first imported into North America in the 1850s, and now over 90% of North American dairy cows are Holstein. With the miracle of artificial insemination — where semen is collected from a bull, frozen, and cryogenically stored — one bull can sire millions of head in his lifetime and beyond. As the genetics companies go global, a very small set of bulls can quickly come to dominate the dairy industry globally.

The beef industry is a little better. No one breed has dominated to the same extent that Holsteins have in the dairy industry. The beef industry, on the other hand, has allowed a slightly varied definition of purebred compared to all other animals. Usually, purebred means that both the parents of an animal are registered purebreds themselves. This means that each breed has a somewhat distinct genetic pool. The beef industry calls this definition "full blood" and allows "upgrades" to occur within the purebred definition. Essentially, a purebred animal can be crossed with any other breed and then returned to purebred status in three generations. For example, an Angus breeder that wants more milk production and a larger carcass size could cross one of their Angus cows with a Simmental bull. Their offspring would be 50% Angus and 50% Simmental. The breeders would then select the best out of their offspring and breed them back Angus. The next generation is 75% Angus. If they do that twice more, their bulls are 15/16ths Angus and can be called purebred with an asterisk on their pedigree indicating they are not full blood. Once their percentages get to 99% Angus, the asterisk disappears.

How important is it that the beef industry can "upgrade" purebreds? I'm not sure. Now that we have the technology to analyze DNA, we can look at populations of animals and analyze how diverse their genetics are or how interrelated they are. As we drive around the countryside, we see what appears to be a very diverse population of beef cattle. The fact is that they are a lot more closely related than their hides make them appear. Researchers were surprised to find that most commercial beef breeds were fairly closely related. The White Parks, which we raise, are one of the few truly distinct gene pools.

The Angus breed brand has come to dominate consumers' perception of quality beef. The thing is, when the Angus program started, the selection for the Angus program occurred in the coolers after the hides were removed. All those "Angus" steaks you've bought could have been from any breed — they were selected for high-quality meat, not their breed. A substantial portion of cattle are still sold to the packers live and priced on live weight. Even cattle buyers shifted their emphasis from their perception of what an animal's carcass would be to include the colour of the hide — a black hide now brings a premium at the stockyards. Angus used to be the only solid black breed. There are now Black Simmental, Black Limousin, Black Charolais, and on down the list. As I've stated, the black hide isn't the only gene being transferred. With all these mash-up crosses, the genetics of all the breeds are moving towards one another.

Why should we worry about declining genetic pools? Sexual reproduction and genetic diversity have been key components in the population of any species adapting to change. Tossing breeds on the scrap pile because they aren't as productive or don't measure up in today's production systems is foolhardy because we don't know what economically important genes we are losing. I'll give you an example: In the sheep industry, we are quickly coming to the end of the chemical parasite control era. Parasites are developing resistance to almost every known pesticide. It's a problem that started in the warmer regions and is spreading rapidly. There are sheep that are genetically more resistant to the parasites than others. It's a trait that hasn't been selected for in the last 50 years because there were effective pesticides to combat the parasites; however, a particular breed of hair sheep — Katahdin — has a resistant genotype that breeders have been selecting for in recent years. It's possible that the resistance gene is also in other breeds, but Katahdins are the first breed with demonstrable resistance. If we had lost that breed, we might have lost natural resistance with it.

In the case of cauliflower resistance to black rot, we did lose these genes for half a century. There were varieties of open-pollinated cauliflower that were resistant to black rot, but French hybrid varieties swept England; in the space of less than a decade, all the heritage varieties had been replaced with the hybrids. The hybrids produced more uniform heads and needed half the nitrogen fertilizer of their open-pollinated counterparts. By the time someone realized that the hybrids were susceptible to black rot, no resistant genetic material could be found. We are just now breeding black rot resistance back into all the cruxifers[46] after finding resistance in a wild mustard population.

AGRICULTURE AND SOCIETY

There is an old saying: "Wisdom is knowledge that you gain immediately after you need it." I'd rephrase a little and suggest that the value of biodiversity doesn't become obvious until after you've lost it and need it once again.

Some will argue that there is nothing to worry about. We have frozen semen from almost every breed of cattle currently in existence. We also have the global seed vault in Norway that is collecting seeds from numerous varieties of food crops from around the world. These are good initiatives, but they're not sufficient. Storing semen only captures the male side of the genetic base. There are genes and genetic information only carried by the female half of the species. There is also an incredible amount of knowledge that is necessary to successfully cultivate each variety and species — not just knowledge of the specific plant or animal but also the ecosystem that surrounds it. Just having the seeds or semen isn't enough.

When we started our transition to grass-fed cattle and sheep, I began by planting hay/pasture. Today, my most productive and resilient pasture is the one that I didn't get around to planting. It was a spelt field,[47] and it had enough clover grow the following spring that I decided to graze it before breaking it up to plant a pasture. After grazing, we were starting to see signs of a drought setting in, so I left it for fear of not getting a good catch with what I intended to plant. The following spring I was going to break it up again — it was a little rough in places, and not all the species are what I would have planted. But it was a wet spring, and by

---

46. Cruxifers are a family of vegetables that includes cabbage, cauliflower, broccoli, bok choi, and many other greens.
47. Spelt is a grain similar to wheat.

the time I could break it up, I couldn't convince my wife that the time and diesel fuel would improve the yield or feed value of the pasture, so I left it.[48]

This pasture has several advantages that let it out yield my other pastures. First, every plant growing in it has successfully grown there in the past. It is not hampered by my perception of what should grow there or would make better cattle feed. Second, I was much further along in my grazing education when it became a pasture than I was with any of my other fields. Thus, this more recent pasture doesn't have to recover from all the mistakes I made with the other pastures. I have the bags of seed that I intended to plant in that pasture still sitting in my shed. The most important thing for a productive pasture wasn't the seed; it was the knowledge of the farmer. As an extra bonus, I have a much more diverse pasture. There are clovers, vetches, and grasses growing that I would never have planted (primarily because I would have had a hard time finding seed to plant.)

Another time I had lots of seed but little knowledge were the years we ventured into the market garden business. I ordered an impressive list of vegetable seed from a number of seed suppliers. I had grown a garden and thousands of acres of corn, soybeans and grain — how tough could a measly five acres of vegetables be? It was plenty tough. Just because a variety or vegetable can grow in your general climatic conditions doesn't mean that it will thrive on your particular farm, or taste very good. By our third year of vegetables, we were starting to develop a list of favourite varieties and species that consistently performed on our farm.[49] It's a good thing we started out with a long, diverse list of varieties and species that we could gradually shorten. That's how agriculture generally works — we gradually reduce diversity to focus on a few high-performing plants or animals that best meet our current needs.

Agriculture takes a hard rap for destroying habitat and endangering the lives of many species, both here in North America and around the world. There is value to those species beyond being part of our natural heritage and aesthetic appeal. I go back to resiliency. On our farm, we raise sheep

---

48. I couldn't convince myself either, but it makes a better story this way and if it goes off the rails I have someone to blame. (I love you, Sil.)
49. We never completed the exercise because we realized that there were other growers who were more passionate about growing vegetables than I was ever going to be. Watching calves and lambs play beats watching lettuce grow by a wide margin in my book; although, the lettuce never knocked down fences and ate lambs or calves.

surrounded by woodlands on three sides. I fall asleep listening to coyotes many nights, and I don't lock up our sheep in the barn at night. How am I able to get away with that? Part of the answer lies with our dogs and various forms of electric fence we have, but I think an equal or maybe more important factor is the diversity of wildlife that is already present in those woodlands. If a coyote is well fed in the bush, it's not going to venture too close to the barns — the meals are just too hard to get when you have to fool a couple of dogs and risk getting an electric jolt.

If the coyote can't get a good meal in the bush, however, then there aren't many barriers that will prevent a coyote from sampling the farm's buffet. I remember a particularly bad winter in my youth. The snow was deep and had made the pickings slim in the wild. A particular coyote/wolf was becoming brazen, and due to the conditions, the farm livestock became the easiest target. He attempted to drag a calf right out of a barn while the farmer was standing there. The neighbours tracked that wolf for almost a week before they caught him. One neighbour now has a very nice wolf-hide rug in front of his fireplace.

The wildlife–agriculture interface is a challenging one to manage. We participated in research looking at the impact of the timing of first cut hay on the successful nesting of bobolinks because they're listed as an endangered species in our area. A conservation group has bought up a large parcel of land about 50 kilometres north of us as a nature reserve to help the populations of the loggerhead shrike and the bobolinks. Initially, they kicked all the grazing cattle off the land. Then they realized that the grazing of the cattle was helping to preserve the grassland habitat of these birds. They now allow a grazing operation to run cattle on their lands to maintain the grasslands and prevent the scrub trees from taking over. The initial assumptions of the conservationists about the impact of agriculture were incorrect.

Some people thought I was foolish in allowing researchers to document an endangered species on my farm. Under new legislation, the government has some fairly onerous tools to protect endangered species on private land. The legislation was written with very little consultation with agriculture — it was the bailiwick of the environmental groups. It remains to be seen whether the government will invoke its new rights. My perspective was that we needed to find a way for these birds and agriculture to coexist. Society shouldn't be able to just expropriate a farmer's land without compensation by imposing restrictions on what can be done, but if we can identify a critical window when hay harvest should be avoided to limit the impact on bobolinks, for example, then let's find it. Interestingly, in our jurisdiction bobolinks are classed as endangered, but when they fly south from here to the US southeast, they are considered a nuisance bird and are often killed to prevent crop damage. So, if there is a good chance somebody somewhere else is going to kill them, should I modify my farming practices to save them? Bobolinks aren't even native to Ontario; they are grassland birds that extended their territory into Ontario after farmers cleared the forests. If bobolinks aren't native to Ontario and are sufficiently numerous to be considered a nuisance in other jurisdictions, do Ontario farmers still have a responsibility to protect them?

From the perspective of many environmentalists, farmers should be leaving fencerows intact and clearing no more land. I understand that. Fencerows are important connecting links between the islands of bush habitat. Woodlands represent such a small percentage of the landscape in agricultural regions that there is a good case to be made for preserving them. Over time, fencerows also help slow both wind and water erosion. I also understand the other side. It costs five to ten times more to buy an acre of farmland than to clear land a farmer already owns. If there is a bush that can be cleared, it's a very cost-effective way to increase productive land base. There are two reasons to clear fencerows: First, they represent land that could be productive. Second, they cause inefficiency. When you take out a fencerow and join two fields, you eliminate a set of "headlands" where you have to lift equipment and turn around. In effect, the headlands have to be covered twice for every operation. We've done the math and you can gain close to 10% in time when working a 100-acre field versus two 50-acre fields. When you're running hard against the weather during planting or harvesting, that time gain can mean a lot.

So how do we, as a society, ensure that the farmer's best interests are aligned with society's? On one hand, there is the concept of private property rights firmly enshrined; but on the other hand, fencerows and woodlands have value to society. A lot of environmental groups see the solution as legislation. Legislation preventing farmers from removing fencerows and clearing bush, however, is akin to expropriation without compensation. Each property would have less value to the farmer after the legislation than before it. How would you like it if the government passes legislation that says you have to turn one room in your house over to some purpose to benefit society, reducing your enjoyment, and

126

possibly the value, of your house? It's easy to talk about imposing society's will on some faceless farmer for the greater good. It's a little different when put into perspective.

Farmers and environmentalists don't have to be adversaries. Alternative Land Use Services (ALUS)[50] is an alternative that some conservation groups are using to work with farmers. There are numerous benefits to this approach. The first one is that the conservationists and the farmers are sitting down together and having a dialogue. The adversarial nature of the relationship is eliminated, and bridges can be built around common purposes. Farmers intuitively understand the value of conserving the environment since our livelihoods are inextricably intertwined with the natural world. Even as we use ever more technology to exert control over the natural environment, we understand that the natural environment has more impact on our livelihoods than we do. Our challenge is to survive in a commodity business while still being able to invest in the protection of the environment.

ALUS programs involve the entire community in the discussion of what the highest priority issues are. Funding is provided to farmers for making environmental improvements to their farms. Projects have included:

- planting tall grass prairie;
- reforestation, including oak savannahs;
- planting windbreaks;
- planting hedgerows with flowering and fruiting shrubs;
- planting shelterbelts along riparian areas; and
- creating land dugouts for storage of water during extreme rain.

These ALUS programs have generated some impressive results. Outcomes have included:

- increased bird populations;
- increased populations of native pollinators;
- reduced stream pollution from eroded soil;

---

50. For more on Alternative Land Use Services (ALUS), visit their website: http://www.alus.ca.

- establishment of more drought-resilient pastures;

- increased protection of land from extreme weather events; and

- increased waterfowl habitat, including successful reproduction.

John MacQuarrie, Deputy Minister of Environment, Energy, and Forestry, PEI, has discussed why he prefers ALUS over regulation:

> *You can't legislate positive environmental outcomes on farmland. It's inefficient economically and sets up a tough dynamic with rural communities. There's nothing worse than dragging a farmer into court.*[51]

ALUS is successful because it involves the entire community, including the private farmland owners. The ALUS project in Norfolk County in Southwestern Ontario, for example, has 78 partner organizations that are supporting the project in one form or another.

It is a challenge to know whether a farmer is watching out for the environment and biodiversity since there is no magic certification stamp. I can be certified organic and still rip out fencerows and push back forests. Most organic farmers are disgusted to learn that rainforest in South America is being bulldozed to plant organic crops in order to compete with us in the international organic grain trade; however, this is the cheapest way to quickly expand organic production and capture the premiums that are available. Clearly, these are people who aren't following the spirit of organic regulation but rather the letter.

While organic agriculture has been proven to improve wildlife diversity,[52] BioSuisse is the only organic certification program that has a wildlife habitat component to its certification rules that I am aware of. Under their rules, non-cropped area has to be a minimum of 7% of a farmer's land base. The Local Food Plus certification program here in Ontario prohibits any clearing of forested land to qualify for their certification. I'm sure there are a patchwork of other organizations that look at this dimension of agriculture; however, the best way to be sure that the food you eat comes from a farm that values biodiversity is to get to know your farmer. Do they share your values? Does their twitter feed include comments on sighting endangered species? Does their website

51. Rob Olson, "Alternative Land Use Services."
52. "Biodiversity and Organic Agriculture," Nadia Scialabba and Caroline Hattam, Food and Agriculture Organization of the United Nations, accessed on July 23, 2013, http://www.fao.org/ organicag/_doc/biodiv_OA.htm.

include pictures of wildlife on their farm? Understanding the impact of your food choices on biodiversity is not easy if your food is produced 1500 miles away or on another continent. It's only when you choose to shorten the distance between you and your farmer that you can make an accurate assessment. Unless more people start caring, more species will be extirpated in their regions — and some will become extinct. You can choose to limit the impact of your food choices on biodiversity.

CHAPTER 10

H 2 UH OH!

We never know the worth of water till the well is dry.

— *Thomas Fuller*

There are a few phrases on our farm that can cause everyone to stop what they're doing and focus on the speaker: "The cattle are on the neighbour's lawn." "The sheep are on Opa's lawn." "There's no water in the barn."

When you have livestock, you are acutely aware of the importance of water. Cattle react to a water shortage in very similar ways to how I expect humans will. If I can restore the water flow before the trough is drained, our cattle carry on as if nothing happened. If the trough goes dry and the cattle become thirsty before I have the problem fixed, an interesting dynamic occurs. The boss cows in the herd will guard the water trough and prevent the lower-ranked cows from drinking. Essentially, they hoard the water supply. If the problem was caught early and hasn't occurred in a while, very little hoarding occurs. If we've had intermittent outages for a few days, however, things can get nasty.

We have contingencies in place — turning a tap switches between two wells that are independently capable of watering the herd. We have two back-up generators if electricity is the problem and, in the absolute worst cases, pond and river access. But lack of water supply has only been our problem once, and we replaced the shallow bored well with a deeper drilled well. The problem usually is related to a leak somewhere. We have three kilometres of surface waterline distributing water to our current pastures, and we probably need another kilometre or two before I can rotationally graze all areas of the farm properly.

It's amazing what can cause a leak in a water line. Sometimes, it's human error — opening a branch line and not checking that all the valves are closed. Other times it's a direct act of malice by an animal. We used to have a 375-litre water trough that was too tall for the cows to reach the absolute bottom without tipping the trough slightly. If two were arguing over whom got the last bit in the bottom of the trough, they would tip it over. If they would have waited five minutes for the trough to refill partially, they could have easily gotten a full drink. Instead, they ended up draining a well and creating a puddle. When you'd arrive to right the trough and let it fill with water, at least one cow would indignantly stand there and bellow at you as if to say, "Don't you think you could move a little faster, I'm thirsty." There's definitely never a "Thank you" once water is restored — more of a "Don't ever let this happen again."

We're gradually replacing the pieces of our watering system that cause the most headaches. Auto shut-off valves are going in all new water lines, and any valve that needs repair is being replaced. The water trough that could be tipped has been replaced by a water trough with shorter sides and a larger diameter that can't be tipped (which directly resulted in my back injury last year when I tried to empty it in a hurry by tipping it. I could tip the old trough when it was almost full. This one needs to be almost empty.) Plastic fittings that snapped easily have been replaced with steel where needed.

The old shallow well had a jet pump, which required "priming" if it ever sucked air. Priming involves pouring water down through a small opening on the top of the pump until you have the supply line sufficiently filled so that the pump will create enough suction with the priming water for the whole system to start flowing again. I remember one day when my children's attitudes on water conservation caused a small kerfuffle. For reasons that I will never understand, the jet pump that supplied the barn and livestock was located in the house that my father-in-law lives in. The closest spigot is 90 metres away in the barn. To prime a pump, you need a spigot open to allow the pump to push water somewhere. This particular pump was always a challenge to prime; however, on this particular day, it was being more temperamental than usual. I finally suspected that someone had closed the spigot. I came up out of my father-in-law's basement, walked to the barn, walked out to the spigot in the pasture that I had opened so the water would just flow out onto the ground, and discovered that it was closed. I stormed to the house and demanded to know why it had been closed. Didn't everybody know I was trying to prime the pump? My middle son (who was probably 11 or 12 years old at the time) piped up with this explanation: "I shut it off because it was just wasting water on the ground." His logic was perfect. Under any other circumstance, I would have expected him to shut off the water if it was spilling on to the ground. In a very controlled voice, I announced to the entire family that I needed that water running to fix the problem with the pump and that I would be responsible for turning it off. When we drilled the new well, it was outfitted with a submersible pump that didn't need priming.

Too much and too little water are both problems for agriculture. We are very fortunate where we farm — in a normal year, we get around 80 centimetres of precipitation more or less evenly distributed over a year. February is our driest month. But, as an old-timer once told me, "A *normal* year is just the average of ten *abnormal* years." Our first year with the market garden had the driest summer in recorded history for our little part of Ontario. We set up some makeshift drip irrigation and made it through the season. We received many accolades for how sweet and tasty our produce was from both chefs and our market customers. The second year of our market garden we received 20 centimetres of rain in June. We didn't start a tractor for the entire month — there was no point, our fields were mud. We dried out somewhat for the balance of the summer, but we ended up with one of the wettest summers in recorded history. Many of our customers commented on our produce being watery and not nearly as sweet. Too much rain and not enough sunshine will do that. We had invested in a full irrigation system and never turned it on that year. When you receive too little rainfall, a farmer can add more water with irrigation. When you receive too much water, there's not much a farmer can do.

This is one of the reasons a significant portion of North American produce is grown in a desert. For all intents and purposes, the central valley of California is a sunny desert, and it's a great place to grow produce. A farmer has almost complete control over the water. When he sets out transplants, the irrigation can be turned on to ensure the plants get a good start. If the plants need more fertilizer, it can be mixed in the irrigation water. If he needs to harvest in a field, he shuts the irrigation off a few days before so the ground isn't wet. It's a wonderful way to farm, unless you're running out of irrigation water. And many areas dependent on irrigation are running out of water. About 70% of water withdrawals globally are used in agriculture, and between 15% and 30% of irrigated water is drawn from unsustainable sources.[53]

However, while agriculture dominates water use globally, every aspect of our civilization is in some way dependent on water. In the foreword to the World Economic Forum's report *Water Security: The water-food-energy-climate nexus*, Margaret Catley-Carlson, Patron, Global Water Partnership, Canada, writes:

> *Water infuses not only our ground beef patty, lettuce, cheese, pickles, onions, ketchup, and sesame seed bun, but also the bag and packaging in which that hamburger is provided, the building in which it was grilled, the energy to cook it, and the financial system that lent the franchise capital. River currents turn turbines or grow fuel or cool plants that generate its electrical currents. On fresh water dangles the life or death of five thousand children each day, the clothing they wear and whether their weak governing state will grow stable or start to unravel.*[54]

UNDERGROUND AQUIFIERS AND DESERTIFICATION

Water is being drawn from underground aquifers significantly faster than they are being replenished across the entire western United States. Most rivers flowing through California have had most of their flow diverted. These are primarily glacier-fed rivers, and those glaciers are shrinking due to climate disruption. Have you ever looked at the Rio Grande River on the border between Texas and Mexico? We drove along it in late January of 2005. It was one of the wettest winters in the US Southwest, and I've seen more water flowing in the drainage ditch across my farm. When we inquired as to why the river was so low, the reason was most of the water was held behind dams up river and then slowly released during the growing season so the farms along the river could irrigate. Humanity has pumped so much water out of underground aquifers without it being replaced that it has had a measureable impact on sea level and the rotation of the earth. There is land on most continents that has sunk because so much water has been sucked out from under it. Obviously, this

---

53. Millennium Ecosystem Assessment, *Ecosystems*.
54. The World Economic Forum Water Initiative, *Water Security*.

can't continue indefinitely. We're already seeing water disagreements in California between farmers, cities, and people trying to prevent the inhabitants of entire ecosystems from being extirpated. These conflicts will only continue to increase.

Not only are we drawing water from aquifers faster than they are replenishing, we are also slowing the rate at which they replenish. When an area is denuded of all trees and vegetation through either mismanagement of livestock or the desperation of its citizens, the removal of the vegetation reduces the amount of water that enters and is retained in the soil. Water that soaks into the soil can do one of three things: evaporate back into the air, be drawn in by plants and transpired, or seep deep into the soil to replenish aquifers. This might seem a little counterintuitive, but water can be absorbed and retained by soil covered in vegetation much more readily than bare ground. As well, evaporation from bare soil is greater than transpiration from vegetation. Thus, on perennially bare ground, water never gets a chance to seep deep into the soil, and the hydrological cycle is gradually destroyed. It's one of the processes by which desertification occurs.[55] As we drove from El Paso across to Tucson, I found it more than a little amusing that you would see a road sign warning of the potential for dust storms followed almost immediately by a sign warning of the potential for flash flooding — sometimes these two signs were on the same post. Destruction from an excess of water and a dearth of water in the same place isn't that surprising given they have their root in the same ecosystem problem.

I'm aware of two projects that are trying to reverse desertification on a large scale. The first is the aforementioned work of Allan Savory. If nothing else, watch his TED talk, titled "How to fight desertification and reverse climate change." He shows before and after pictures of the restoration work that has been accomplished with holistic planned grazing of cattle. No expensive inputs, machinery, or patented seeds is needed; it's just a new way of managing the interaction between large herbivores and the landscape — a case of biomimicry that is working exceedingly well.

The restoration effort for the Loess Plateau in China is an example of a completely different approach to restoring an entire watershed. In this case, the government of China, the World Bank, and several other

---

55. Desertification is the process by which generally arid land is degraded to the point that it's hydrological cycle is destroyed, leaving it devoid of bodies of water, springs, and their associated vegetation.

organizations invested over a $100 million to re-establish the hydrological cycle in an area that had been the breadbasket of the Hahn Dynasty but has barely supported subsistence agriculture for the last millennium. Their methods involve reshaping the landscape with large machinery to capture and retain water, paying residents to plant an incredible numbers of trees, and introducing a form of private property rights to encourage people to properly manage the land. The transformation has been considerable. What was once a barren wasteland is now covered by lush vegetation. Whether the transformation is permanent will depend on the decisions citizens of the region make in the next centuries.

These are examples of the positive impact of humanity interfering in "natural" cycles. The Aral Sea is an example from the other end of the spectrum. In 1960, the Aral Sea was either the third- or fourth-largest inland lake by surface area depending on whether the Caspian Sea was counted. Today, for all intents and purposes, the Aral Sea does not exist. The massive irrigation projects constructed on the two major rivers flowing into the Aral Sea diverted so much water that the Aral Sea evaporated — literally. It has no outlet; the only water that leaves the Aral Sea leaves by evaporation. Today, a toxic salt flat exists where the large sea once was. A thriving fishing industry, as well as the local economy, has been destroyed. The moderating impact of the sea on weather patterns in the region has been lost. The irrigation projects were started by the Soviet Union. Today, the Aral Sea forms part of the border between Uzbekistan and Kazakhstan, and four additional countries have portions of the watersheds of the rivers flowing into the sea within their borders. The politics of re-establishing the water flows on the rivers and allowing the Aral Sea to regenerate will be almost impossible to navigate.

WATER RECYCLING

Water can be an infinitely renewable resource. Ecosystem processes and services can completely clean and recycle water indefinitely; however, we learned a long time ago that the ecosystem processes aren't sufficient to deal with either the nature or quantity of contaminants that human activity generates. We established water treatment processes and standards to limit the pollution flowing out of our sewers. Similarly, air pollution regulations have been introduced to limit the pollution falling from the skies in rainwater.

What we've been much less successful in reducing is the quantity of water we "consume".[56] The water we directly consume in our houses is obvious but actually a minor portion of both the consumption and the waste. Most cities waste the rain — impervious surfaces direct the water to storm sewers and it is immediately discharged without having a chance to soak in, nurture plant life, and recharge the soil and ground water. When we drain wetlands and clear forests, we are wasting rain. Both ecosystems trap water, help the infiltration of water, and moderate the impact of rains on downstream areas. As climate disruption increases the frequency of large rain events, the ability of our landscape to retain and slowly release that water will be paramount to reducing the impact on not only our built infrastructure but also the water cycle itself.

Agriculture has reduced the water wasted by irrigation by an order of magnitude in high-value crops by switching from gun-style irrigation, which blasts the water into the air, to buried drip tape, which releases the water in the root zone. Losses to evaporation are almost eliminated, as is water loss due to applying water between the rows where it is not needed. Some progress has been made on overhead, centre-pivot irrigation systems for lower-value crops by changing the design of the emitters and water pressure to increase droplet size, thereby reducing evaporation losses. However, agriculture needs a paradigm shift in water consumption. Instead of depleting water resources to artificially sustain crops in places they wouldn't normally grow, we need to think about growing crops that can be sustained by the natural water cycle. Bare ground annual agriculture is the single most wasteful crop production system imaginable. If we asked engineers to design a system of food production that wastes the maximum amount of water possible, I doubt they could come up with anything better. The evaporation that occurs during soil preparation and planting up until the crop has grown sufficiently to provide full ground shading is enormous. With my perennial pastures, I can generally be harvesting (grazing cattle) before the majority of annual crops in our

---

56. I give "consume" the broadest definition possible. I include water that is taken from surface and ground water sources to be used in some way and water that is polluted by human activity.

area are planted. While crop farmers are waiting for the water to evaporate and the soils to dry out, I am growing a crop and allowing water to seep deep into the soil. While this distinction isn't devastating in our humid region with relatively uniform rain patterns, it is critical in areas of dry land farming. What's most interesting to me is that the same methods that protect soil and reduce water pollution from farms also improve water use efficiency.

Cities have been less successful in reducing water waste. In fact, we need a new paradigm for city consumption of water. The level of water consumption by most cities stresses the capacity of the surrounding region to provide sufficient water, and a majority of the water "consumed" by cities could be recycled. I'm not talking about grey-water diversion.[57] I'm talking about hooking outlets from sewage treatment plants to cities' water intake pipes. Radical, I know. (People might be a little more careful with what they flush down the toilet if they know they'll be drinking that water again in a few days.) We have the technology to do it today. If the thought of drinking water that has already been through a set of kidneys disgusts you, I hope you don't live in a city or town that draws its water from a river downstream of another town (for example, anywhere south of Minneapolis on the Mississippi). You are consuming water that came out of the upstream waste treatment facility. It will be diluted, but it will be there every time you turn the tap on. And don't go reaching for a bottled beverage if you are now disgusted by the thought of drinking town water. Most of the water in beverage bottles (including the pure water with the glaciers and bucolic springs on the labels) comes from municipal water supplies. Your cheapest source of water — tap water — is also the most regulated and safest water supply available to you. The testing regimes mandated for municipal water supplies are unequalled in the beverage industry. By drinking tap water, you can increase your water consumption and decrease its impact on your budget (not to mention the health benefits from drinking water instead of other beverages).

Many of you in coastal regions will point to desalination of ocean water as an alternative, but desalination has several problems. First, if evaporation is used to purify the water, it requires a material amount of energy. Alternatively, if reverse osmosis is used, significantly more

---

57. Grey water is household waste water from all sources except the toilet. Water containing human waste is generally referred to as black water.

water needs to be drawn in than is produced. Finally, regardless of method, there is a waste water stream created that contains all of the salt and other contaminants removed from the water in the desalination process. Desalination is an expensive diversion that will be a short-term solution at best. The unnecessary level of energy consumption and amount of pollution created mean that desalination is unaffordable in the long run. Matching our use of water to the ability of the ecosystem to supply it is the least cost solution long term. As soon as we start to extract, move, or purify water, the cost increases. It is in our own economic best interest to limit our use and maintain the purity of our water supplies.

I hope that science and reality will prevail over politics and prejudice. Unfortunately, I doubt that will be the case. I can't imagine any sane politician (assuming such a person exists) running on a platform that includes having constituents drink recycled sewer water, regardless of how much economic or environmental benefit would be created. The only way to create true change in society is for public opinion to change. The people generally lead the politicians. Similarly, I doubt water usage will be curbed much in cities or agriculture drawing ground water until the general populace is directly impacted. In North America, it is still a big deal if people are forced to limit the watering of their lawns and the washing of their cars.

The challenge will be finding a mechanism to ration water usage. My economist side says the solution is obvious — put a price on water. But then my farmer side kicks in and says, "You'll put a meter on my well over your cold, dead body…if they ever find it." Agriculture consumes such a large portion of water that any solution that doesn't include agriculture will have limited impact.

OUR FRESH WATER RESERVES

In parallel with direct consumption, we seem to have a blatant disregard for protecting the purity of our fresh water reserves, especially if the choice involves fossil fuel production. And I'm not just talking about the huge spills that garner international attention. In the Canadian province of Alberta, there has been an average of two crude oil spills from pipelines per day for the last 37 years — and that average only includes spills from provincially regulated pipelines and spills large enough to be reported.[58] It doesn't include spills at wells, spills at refineries, or spills of hydrocarbons other than crude oil. I have to have containment around my diesel tank "in case of a spill." Yet a spill of the entire contents of my tank wouldn't be large enough to be reportable if I owned a pipeline. Wouldn't a series of small spills by a pipeline operator be an indicator that there is an increased risk of a big spill?

Agriculture has to share some of the blame for fresh water contamination. Atrazine and other agricultural pesticides have contaminated many groundwater sources. A portion of the turbidity and pollution in any major river system globally originates with runoff from agriculture. Large spills of liquid animal manure have polluted rivers: the increased levels of phosphorus and nitrogen in waterways originate with agriculture.

At the Stoddart Family Farm, we are attempting to minimize our impact on fresh water resources. We ensure maximum water infiltration by having the majority of our land in well-managed perennial pasture. It's good for the environment and our bottom line. Last year, during the major North American drought, our rotational grazing allowed us to continue grazing throughout the drought without feeding any hay or losing animal condition. Other farmers in our area using traditional grazing fed hay through most of July and part of August. Our vet was out to do some pregnancy checks in November and commented on the good condition our cattle were in. We were still grazing at that point. He had some clients whose cattle were in poor condition despite being fed hay. Capturing more water and losing less of it to evaporation is important to our long-term financial viability.

Increased water infiltration means decreased runoff and potential for pollution. Perennial pastures also prevent erosion and soil contamination of waterways. Our land is certified organic, meaning we aren't applying any pesticides or synthetic fertilizers that could potentially become pollutants. Protecting water quality isn't just about protecting the environment; it's also self-serving. Our drinking water comes from groundwater under our farm, and our favourite swimming hole is downstream from our farm.

Maintaining good water infiltration is also important to the health of a community. We are fortunate to live in an area with abundant and

---

58. About half the pipelines in Alberta are regulated federally because they cross a provincial or international border.

consistent rainfall — at least currently. Recharge of the groundwater isn't really an issue. In our area, most of the water problems are on the other end of the spectrum, where tiling and ditching are done to remove water and allow land to dry out sooner in the spring so that we can work it. Whether that continues in the face of climate disruption is anyone's guess. I do know that water is critical to our farm and that vibrant ecosystems and economies can collapse quickly when the water cycle is destroyed.

How do you know if the food you're eating was grown in a way that enhanced or diminished the water cycle? There are very few labels that you can watch for. "Product of California" means there is a good chance the produce was grown in a desert with unsustainable water use. The organic rules don't provide much guidance on water use other than testing the source of irrigation and the livestock water for contaminants on a regular basis. If you're shopping at a local farm, it's much easier to tell. If there are irrigation guns spraying water into the air, water is being wasted. If the ground is kept bare for long periods during the growing season, water is being wasted. If the cattle and sheep are not out on pasture, water is being wasted. Yes, rotational grazing reduces water consumption. Living green plants provide a material portion of animal's water requirements when they are out on pasture. I'm always surprised how much water intake increases when we start feeding dry hay. Besides, the water harvested in plants on pasture is rainwater, and the water supplied in drinking troughs is groundwater. If I ran a feedlot with the same number of cattle, I would use a lot more groundwater and would put pressure on the ability of the aquifer to naturally recharge.

This even holds true when we are grazing stockpiled forages in the winter with snow on the ground. The cows meet a lot of their water consumption needs by consuming snow. I know at least a few graziers that rely 100% on snow to supply water needs when grazing in the winter. (It sure beats thawing frozen waterlines.) I had heard of the practice but never completely bought into it until I saw it at work on our farm by choice of the cattle. Two winters ago, I did some winter grazing on a field that was a good distance from the barn. I left gates open between the barn and the pasture so that the cattle could come back for a drink from our ice-free water troughs in the corral. One day, I noticed that there were no cattle tracks up to the barn in snow that had fallen several days previous. I could see that the cattle were all contently grazing or lying around chewing their cud. They clearly were getting what they needed without coming to the barn.

Even in Canada and the US, where household water use is well above the average consumption anywhere else, agriculture consumes more water than municipalities. Your food choices can have as large an impact on the water footprint of your lifestyle as your direct water consumption choices.

CHAPTER 11

HALF BAKED

You have much more power when you are working for the right thing than when you are working against the wrong thing. And, of course, if the right thing is established wrong things will fade away of their own accord.

— *Peace Pilgrim*

A few years ago our youngest son, Harrison, was working our farmers' market booth with us. He was eight or nine. A 20-something customer walked up to our stall, picked up a vegetable, and asked Harrison what it was. Harrison told him. The customer then asked him how to cook it. Harrison gave him several suggestions. The customer thanked Harrison and put the potato back in the basket. Several ladies in line were quite amused that our youngest son had just given accurate cooking instructions to a 20-something adult. It's an amusing anecdote, but it's also a sad commentary on the cooking skills of a generation. Assuming the customer had eaten potatoes out of his mother's kitchen, he must have only seen instant mashed potatoes or ready-to-bake fries. If you don't cook but only reheat processed food, you are putting your diet and health in the hands of someone else.

Imagine that you walk in the house and you can smell brownies baking. You see organic fair trade chocolate and organic flour on the counter. The brownies are being baked with all the finest ingredients. You wait impatiently for the brownies to finish cooking. Finally, the timer goes off, the chef opens the oven, and the tantalizing vapours flood into the kitchen and fill your senses. Your mouth is watering in anticipation of sinking your teeth into a brownie. The chef sets the brownies on a cooling rack. Too impatient to let them cool, you grab one up and take a bite. The chef then says, "Oh, by the way, there was sort of this accident while I was mixing the brownie batter. It's hard to explain, but cat shit from the litter box ended up in the batter. But it was only for like five seconds, and I'm sure that I got it all out. Sorry!" Do you swallow?

From where I'm sitting, there is a whole lot of stuff worse than cat shit that we blindly swallow in our food every day. This coming from a guy who has literally had a mouthful of bull shit and a mouthful of pig shit on separate occasions. Did it gross me out? Hell yeah. Did it start me on my way to insulin resistance and diabetes? No. Did it increase my risk of a long list of diseases? No. But did it increase my risk of never being kissed? Yes. I waited until after Silvia and I were married to share those particular anecdotes with my lovely wife. The image is kind of a mood killer when you're on first base trying to steal second.

Some of you might object to my statement that a minor amount of cat shit baked into a brownie is not as bad as some of the additives that you accept in your food every day. First, let's think about the risk from that cat shit. Other than it being a disgusting thought, the only real risk it presents is microbial, and the baking process will have killed any little monsters. The fact is that people pay a high premium for a beverage that is brewed from cat shit — Kopi Luwak.[59]

FOOD ADDITIVES

One of the most common food additives has been a known contributor to cardiovascular disease since the 1960s. A recent study demonstrated that the 20% of women with the highest consumption of this additive doubled their risk of coronary heart disease.[60] Studies have also shown an inverse relationship between consumption of this food additive and HDL (good) cholesterol — the more you consume, the lower your levels of HDL.[61] Metabolizing it can leach B, C, and D vitamins, as well as calcium, phosphorous, iron, selenium, and zinc from our teeth, bones, and tissues. The CDC found a strong correlation between consumption of beverages containing this additive and obesity. Not surprisingly, there is also a correlation between consumption of this additive and Type 2 diabetes in

---

59. Kopi Luwak, also known as civet coffee, is considered the most expensive coffee in the world. These coffee beans are first ingested and then excreted by the Asian palm civet, a feline native to South and Southeast Asia, primarily Indonesia.
60. S. Liu et al., "Prospective study."
61. N. Ernst et al., "Plasma high-density lipoprotein cholesterol"; Barbara V. Howard and Judith Wylie-Rosett, "Sugar and Cardiovascular Disease"; "Nutrient Intake."

adults. The Centers for Disease Control and Prevention (CDC) in the US recommends reducing or eliminating consumption of this additive in beverages.[62] Despite all of this information, consumption of this additive has been steadily increasing. There are no rules prohibiting the use of this additive in organic foods as long as it is from a certified organic source.

The link between another common food additive and elevated blood pressure has been known for over a century. Elevated blood pressure leads to increased risk of cardiovascular morbidity and mortality.[63] The CDC flatly states that Americans consume too much of this additive;[64] yet consumption is steadily increasing, and the average American diet has over double the acceptable daily intake and over 50% more than the recommended upper limit (the highest amount that can be consumed without risking adverse health effects). Neither of these additives is eliminated or even reduced by choosing organic processed products over standard products.

I hope most of you recognized the two additives I was talking about: sugar and salt. I think they can be properly classed as food additives because they don't provide any nutritional value. Sugar provides empty calories, and salt has no impact beyond a very small level needed for proper body function. (Our diets, however, are so far beyond that minimum maintenance level that we're nowhere near the risk of having too little salt.) There are some really simple ways to reduce intake of both of these additives. The quickest place to reduce sugar intake is in the beverage department — swap out sweetened beverages for water. Your dental bill will go down as well as your risk of Type 2 diabetes. For salt, stop eating processed foods and restaurant meals. Salt is present in much higher quantities in prepared food than it is in home-cooked food because adding salt is a cheap way to make something taste better.

Just so you clearly understand me. Don't swap your regular can of soda for a certified organic can of soda[65] — swap it for water. Don't swap your regular zip-and-zap meal for an organic zip-and-zap meal — swap it for home cooking. Choosing certified organic is only one

62. "Rethink Your Drink," Centers of Disease Control and Prevention, last update June 11, 2013, http://www.cdc.gov/healthyweight/healthy_eating/drinks.html.
63. Michael H. Alderman, "Salt."
64. "Most Americans Should Consume Less Sodium," Centers of Disease Control and Prevention, last update June 19, 2013, http://www.cdc.gov/salt/.
65. Yes, they exist. I can't believe that anyone who is watching their health enough to be buying organic will actually buy organic soda — both regular and organic sodas are carbonated water plus sugar and flavour.

dimension of improving your food choices. When you choose to cook, you can and will choose to lower your salt and sugar intake.

THE COOKING TABOO

I think most cookbooks overcomplicate cooking. They provide exotic recipes to show off with at dinner parties, but fall short in everyday cooking for a family. A good old-fashioned roasting pan in a 245–250°F oven can do wonders without any supervision. If your oven has time bake, you can fill the pan with ingredients in the morning before you go to work, set the timer, and come home to a lovely, perfectly cooked dish. It also makes beautiful work of some of the standard cuts that are difficult to cook quickly without having them go tough. Pot roasts, stewing meat, marinating steaks, pork shoulders all come out succulent and tender. You don't need to spring for a crock pot. (Silvia rarely uses ours.)

I also think food companies have convinced us that we are incapable of cooking good-tasting food without them and their secret sauces. That's a crock of stew. Food doesn't have to have all kinds of exotic flavours and ingredients to taste good. My favourite meal is a roast of anything, mashed potatoes, gravy made from the drippings, and a salad or steamed vegetables. You get to experience the flavours of each part of the meal without it being overpowered by fancy sauces or complicated processes. I've been able to put that meal on the table since I was about 12 years old. If I'm feeling particularly ambitious, the vegetables might come with a nice, from-scratch cheese sauce (if the vegetables are Brussels sprouts, boiled onions, asparagus, or cauliflower then the cheese sauce is mandatory). Don't get me wrong, I love a good pork souvlaki, chicken Parmesan, or Thai satay as much as the next guy. However, a good roast is so simple, so tasty, and so versatile that you will never regret taking the time to learn how to cook one properly. You won't regret it.[66]

---

66. Silvia just brought me a sample of one of our grass-fed short-rib roasts that she slow cooked at 245°F this afternoon. It was melt-in-your-mouth, fall-apart-at-the-touch-of-a-fork fantastic. She took it took it out of the freezer yesterday, thawed it in the refrigerator, and then put it in the oven (if you're short of time, you can go straight from the freezer to the oven). If there's any of that roast left, the guys will make sandwiches from it for their school lunches. You don't need a year at Le Cordon Bleu chef school to replicate that.

SEASONAL EATING

If there's anything that has served to destroy local food systems more than our movement away from home cooking and seasonal eating, I don't know what it is. Eating seasonally immediately conjures up images of cabbages, parsnips, potatoes, carrots, and onions from November to April. But this morning it was minus 22°C, and I had Ontario peaches over waffles made from our own spelt and eggs and a yogurt smoothie made with Ontario strawberries. The root of the issue is really that we need to buy seasonally. The peaches in my breakfast were purchased by the bushel in August and canned for consumption in the winter. The strawberries were harvested from our own patch and then frozen. Did we have fresh peaches and strawberries? No, but they were local products, preserved at the peak of ripeness. There is nothing in the supermarket in January that compares in flavour or cost. Fresh Californian or Chilean strawberries are a ridiculous price, especially compared to pick-your-own strawberries in June. While flavour and texture of transported-in produce have improved over the years, nothing that is picked partially ripe and shipped chilled across a continent can compare to the memory of a sun-warmed, freshly picked strawberry evoked when you crack open a jar of local strawberries you preserved yourself. Similarly, there is no comparison between the industrially canned peach and the one that you canned personally.

Seasonal buying for fresh fruits and vegetables is obvious — it's not hard to know when certain vegetables are in season locally. What is less obvious is that there is also a season for most meat. Industrial agriculture has created models to allow us to harvest meat every day of the year, but many of the mechanisms to do so go against the natural cycle of the seasons. If you are interested in the healthiest meats that come from pasture-raised animals living in a temperate climate, however, there is a season when these animals are at their optimum size, finish, and flavour— late summer and early fall. This means purchasing a large quantity and storing it frozen for consumption later in the year.[67] The bottom line is that eating locally in January doesn't have to be limited to cabbages and

---

67. We encounter a number of customers that have an aversion to frozen meat. Freezing does cause deterioration of meat relative to fresh; however, the faster meat is frozen, the smaller the ice crystal formed and the lesser the damage done. Here's a hint: The flash freezing that butcher shops do is far superior to anything a domestic freezer will accomplish. So if you are going to purchase meat to freeze, you are much better off buying it frozen.

potatoes. If you perfect your seasonal buying strategies and learn some basic preservation techniques, you can increase the local content in your diet, improve the quality of your diet, and lower your total grocery bill.

I wonder if people are intimidated by the thought of trying to preserve food. I learned my preserving skills at the feet of my mother. If I remember correctly, my first job was turning the crank on the meat grinder and feeding in pieces of cucumbers, peppers, and onions to grind them up to make relish — Mrs. Brown's hamburger relish. (I'd been to a birthday party at my friend Rodney Brown's house and his mother had served a homemade relish to put on our hamburgers. I liked it and so I came home with the recipe. It's still in my mother's handwritten recipe book under the heading "Mrs. Brown's Relish.") From there I moved on to cold packing cucumbers ready for dill pickles, blanching beans and packing them in freezer bags, slicing peaches, mashing fruit for jam, and finally, stirring the magical vats of relish and jam while standing on a chair. By the time I was 12 or 13, I could do jam or pickles solo, and I never came close to killing my family.

This year I had a spare Saturday, and so the kids and I picked several quarts of blackcurrants and gooseberries from the bushes we planted a few years ago and then made jam. My youngest measured the sugar (she even did the math on doubling the recipe, in her head). All of us worked at stemming the fruit. We added some frozen rhubarb to bulk up the fruit to the quantity needed for two double batches. We took turns stirring the fragrant pots. This year the family has decided that their favourite jam is "Daddy's," I'm not sure if it's the taste or the memory of making it together.

My point in relating these stories is that once you understand a few basic principles, preserving food isn't rocket science. It's actually a very rewarding family activity. The recipes in the pectin packets are simple — fruit plus sugar plus pectin. I prefer recipes that showcase the flavours of the fruit and have very short ingredient lists, but if you like the more exotic, there are lots of resources on the Internet. Your first batch will be your toughest, but you'll come up the learning curve very quickly. The worst thing that can happen with a batch of jam is it doesn't quite set and you end up with a fruit sauce instead of jam — a fruit sauce that will be delicious over ice cream.[68] There's a foolproof way to know whether

---

68. Actually the worst thing is if you don't stir it while it's at a rolling boil, causing it to stick and burn.

your jars have sealed properly; the tops will suck in and make a beautiful "popping" sound. If a jar doesn't seal properly, stick it in the fridge and consume it in the next few weeks. Consider it an unfortunate reward for your efforts. If you take something out of the pantry to eat and the seal is broken, don't eat it.

If you're considering doing some home preserving, plan ahead. Understand when each fruit or vegetable is most available. You won't find the best deals on fruit and vegetables at your local grocery store; you need to go directly to the farmers. They'll be able to tell you what the best varieties are for preserving — usually the late season varieties. Pick-your-own farms will be the most cost effective and give you a nice family outing as well.[69] If you want to buy the produce at your local farmers' market, make arrangements with a farmer in advance and then make sure you show up. It's awfully tough to move several extra bushels of something late in the day when you realize that the purchaser has forgotten to come.

EATING HEALTHIER AT THE SAME PRICE POINT

"But, Harry, grass-fed and organic are so expensive, how can I afford to feed my family?" If you're a middle-class citizen of any developed country on this planet, you are paying the smallest portion of your income on food of any nation at any time in the history of the world. If you're in North America, you're paying 50% less as a percentage of disposal income than your parents did. But I'll set that aside because as humans we experience everything relatively, and relative to conventional food prices, there is no doubt that grass-fed and organic have higher price points. (If there is a flat-screen TV on your wall, however, you have enough income to cover the increment.)

That said, if you're an average middle-class citizen of any developed country on this planet, I'm willing to bet you that I can switch you to 100% grass-fed and organic and not increase your total food bill. How can I be so bold as to suggest that? One of our customers did it. Her husband was a sceptic. She kept track of their food costs for six months before the switch and then for six months after the switch. She was

---

69. If you take the kids, make sure they understand that they are there to work. Any strawberry consumed before you are weighed out is theft. Farmers know that fresh strawberries are absolutely irresistible straight off the vine, but if your odd sample turns into "one for the basket, one for me," don't be surprised if something is said.

meticulous; she even included the cost of driving out to our farm to shop. In the end, the result was a wash. Her family wasn't forced to eat a steady diet of rice, garbanzo beans, and raw granola (not that there's anything wrong with those, other than maybe the garbanzo beans).

To follow in her footsteps will require some changes:

You'll likely have to change one appliance. You will need a chest freezer to buy and freeze local fruits and vegetables when they are in season. Everything else can be done with your existing appliances (assuming they include a standard range, oven, and a decent food processor). Some of you will also prefer to add a slow cooker instead of doing low-temperature roasting in the oven. Some of you will consider adding a bread machine (Be forewarned: Most of them are cheap pieces of crap that are built to look good on your counter, but quickly break under daily use.) With our family, we need more than one loaf at a time. We have a food processor that does the hard work, and then we rise and bake the bread in the oven.

You'll have to change your buying habits. Forget about fresh strawberries from exotic places in January. You'll have to stop cherry picking only premium cuts of meat. The best deal on meat involves buying a whole side or quarter direct from a farmer, putting it in your freezer, and then gradually eating it. Alternatively, some farmers are offering monthly share programs and buying clubs that get you to price points that are a little more than buying a side but don't force you to put the whole thing in your freezer at once. Community Supported Agriculture (CSA) programs are another way to reduce the cost of your vegetables and meat. You give up control over what you're eating each week, but you're guaranteed the best the farmer has to offer any given week at a reasonable price. Most CSA programs offer drop points in and around urban areas so you may not have to go very far to pick up your farm-fresh produce. There is also a resurgence in food co-ops. A little online research should reveal several options for purchasing direct from farmers.

You'll have to change your cooking habits. If the primary recipe you use begins with "Tear open the packet and microwave on high for..." then you might need a few recipe books or cooking classes. But don't go in for the fancy, impress-your-friends-at-a-dinner-party class or recipe book. Find something that is going to teach you how to efficiently cook healthy meals from regular ingredients.

You'll have to change your family outings centred on food. You'll need to divert some of the dollars spent on dinners out to procuring the ingredients for your home-cooked meals. Family food outings can switch from fast and casual to going to pick-your-own strawberry farms or the farmers' market.

Your cooking vocabulary is going to have to change. You are going to have to familiarize yourself with terms such as blanche, sear, stew, and pressure can.

Last, your purchasing habits will have to change. You get a lot more potato per dollar if you buy a 25-kilogram bag of potatoes in the fall and direct from a farmer than if you pick out the potatoes from the produce display at your local grocery store.[70] Most houses have a cool, dark corner where those potatoes will keep just fine. The same goes for grains and rice. Food co-ops and health food stores can usually source 25-kilogram bags of most rice and grains at a substantially reduced price per kilo. A bushel of peaches bought in season and canned will look like a very good deal in January compared to the tasteless "fresh" fruit imported from halfway around the globe.

If you're concerned about the price of an organic prime rib, I can solve your problem quite easily with one change. But first, I'll explain why it looks so distorted: First, the proportion of people with high-disposable income that shop for organic meat is higher than the proportion in the general population. Second, each animal only comes with so much prime rib, so much tenderloin, so much chuck, and so much blade. A butcher buys a whole or a half animal and has to sell the whole thing before he buys the next one. Since people with higher-disposable income tend to favour the more expensive cuts, the organic butcher has to make them even more expensive to get some people to switch to the lesser known but better tasting (in my not-so-humble opinion) cuts. If you want to lower your organic meat bill, start buying the lower-priced cuts. I'll put a cross-rib roast up against a prime rib any day (as long as the cook knows what they're doing). It's one of the reasons we can make our CSA model work at a fairly low price point. We don't have to price the premium steaks at astronomical prices compared to stew meat to make sure our stock keeps turning. We manage to move whole animals by ensuring everybody gets some

---

70. As an extra benefit, you will avoid eating potatoes that have been sprayed with a sprout inhibitor. Sprout inhibitors prevent the eyes from starting to grow. It's strictly a cosmetic issue. If your potatoes start to sprout, knock the sprouts off and eat them.

premium cuts and everybody gets a fair share of the standard cuts. You will get some high-end steak or tenderloin, but you'll get enough stew, ground meat, and pot roasts to balance things out. But you don't have to sign up for a CSA program to make the same math work in your house. Start buying the less expensive cuts. Talk to your butcher about what you're buying and how to cook it. You'll be pleasantly surprised by the quality of eating experience you will get from cuts that you've never heard of before.

Learning to cook is the best thing you can do for your health and your food budget. If you're still sceptical, I'll explain what the real game in the mass-market food industry is — profit. The easiest way to make a profit in the food industry is to take something low cost that most people wouldn't ordinarily eat (say cat shit) and add just enough flavour and marketing (fine chocolate) to sell it for a premium. My favourite example is instant macaroni and cheese (cat shit) with added broccoli and cauliflower (fine chocolate) to convince consumers that the instant mac and cheese is as good as feeding their kids vegetables. "But Harry, enhancing the nutrition in our children's diets is a good thing!" Well, I guess if you were going to feed them cat shit anyway, a little fine chocolate won't diminish the experience. I taught my kids to eat their vegetables. Hell, I've even caught them stealing broccoli, beans, and tomatoes out of our garden. (They don't know that I know this, but I'd hate to discourage the activity. Although, I was disappointed when I noticed most of the red currants had disappeared.) The veggies are hard to resist when you planted the seeds, watched them grow all season, and have been impatiently waiting for them to ripen.

I once heard a speaker talking about making hay. He said the hay has its highest level of nutrition at the moment the knife cuts it. Everything we do to it after that degrades it. The same is true of most foods. Less processing equals more nutrition. Bottom line: Your diet will be healthier and you will be able to afford higher quality when you cook for yourself. Don't think of it as expensive food; think of it as cheap medicine.

CHAPTER 12

THE ROAD TO DAMASCUS

Lasting change happens when people see for themselves that a different way of life is more fulfilling than their present one.

— *Eknath Easwaran*

Conversion of the food system to truly sustainable won't happen without effort and an external force. Only two per cent of farms in North America are certified organic, and even fewer are what I would call truly sustainable.

You might be wondering why so few have switched if yields can be comparable and there is currently a premium for organic. The issue to farmers is yields are comparable in long-term studies, not in the short term. Research shows you can get back to about 90% of conventional yields after five years in certified organic production, with comparable yields coming in the longer term. It's getting from year zero to year five that is the challenge. The first two years you have to use organic production methods and sell at conventional prices, with yields dropping significantly in the first few years. The transition years are a break-even proposition in the absolute best case. Most of us end up "investing" many dollars per acre to make the switch without mortgage and equipment payments changing. Our family's food bill doesn't change either.

To put the conversion in the context of an urban wage earner, imagine you're coasting along at a reasonable salary. You may or may not see your particular job dying out, but to switch would mean giving up the salary you are earning, going back to school full-time for several years, and then getting an entry level job in your new profession. The upside in the new job is much better than your current position, but you've got mortgage and car payments, and your teenagers have orthodontics. Not many people switch careers until they are forced to by external forces.

The external force that drove me to switch was near bankruptcy. In hindsight, it was akin to Doug Flutie's infamous Hail Mary pass for Boston College. I was fortunate to have rented 100 acres that I could certify immediately. I underwrote the conversion costs on the rest of my land partially from the profits on the first acres and partially with my off-farm salary. From a financial standpoint, we got through it, but there was a definite cost. For six weeks in May and June and from August to December, my evenings and weekends were dedicated to farm work. I'd come home after a full day at the office and an hour commute, change my clothes, climb into the cab of some piece of machinery, and work until conditions wouldn't allow it or I was too sleepy to continue.

One year, I booked two weeks of vacation time in May when we usually got most of our planting done and then watched it rain. Once I was back at work, it would sort of dry up enough to work for 24 or 36 hours before another rain shower would come through. I was lucky in that I was working at a client site that allowed me to take the commuter train into Toronto. The two-hour commute on the train home was all the sleep I got many days that spring. Some nights I came home from the fields for a couple of hours sleep. Some days I came home in time to shower and have breakfast before Silvia took me to the train station, where I would climb into the train and catch some Zs. I was fortunate that the train I rode started at the station I got on at and ended at the station I got off at. The commotion of everyone getting off invariably woke me up. (Although one time on the trip home the conductor found me asleep in a car as he did his final train check after parking the train on a siding for the night. Silvia was sitting in the parking lot wondering what to do as I sheepishly walked back to the station.)

Through all of this, the kicker was that spring planting conditions were so challenging that we planted into a lot of ground that really wasn't ready. It was obvious in the yields we harvested.[71] We would have been better off leaving the tractors in the shed and going fishing. The crop insurance program has an "unseeded" acreage benefit that basically pays you out the rent on your land if conditions were so bad you couldn't get the crop in. The catch is that you have to keep trying to plant as long as they tell you to. If all your neighbours get their crop in, you aren't able to collect on the unseeded acreage benefit. It has to be a regional problem. So we fought the mud and went around wet holes and planted seeds into ground that was too wet hoping that we had made as much progress as our neighbours.

---

71. The technical term is we "mudded" the crops in.

One thing that helped me through those years was my business partner. During harvest, if the window was getting narrow, we would run the combine all night, each of us taking catnaps in the grain truck between loads. There were many months where I spent more time with him than with my wife and kids.

The challenge with juggling a career and farming is that you can't schedule farming. When setting up meetings in May and June, if the client was aware of my farming, I would jokingly ask whether the second rainy Tuesday would work— only half jokingly. There were many days I spent in an office while glorious sunshine was streaming through the windows. It was the price I was willing to pay to continue to farm.

Not everyone is as fortunate as I was. I had a business partner that shouldered the load when I couldn't, just to be neighbourly. I reported to a partner at the accounting firm who was a farm girl and was sympathetic to my situation; although, I made her job easy by making sure I always delivered. I have a wife that does chores when I have to travel. That first summer we still had a full complement of pigs in the barn and I was twice sent for a full week of training. She picked up whatever needed to be done. She was slogging it in the hog barn and raising three boys under the age of five while I was sitting in an air-conditioned, five-star resort in Houston consuming entirely too much at splendid seafood buffets. Yeah, I was fortunate.

I'm telling you all this because you need to understand that the switch to organic wasn't some magical event in our lives where we all sat around, held hands, and sang Kumbaya. Changing production systems isn't like changing crops. It requires an investment in new knowledge and a new mindset.

There are those who believe that the government should institute some kind of program to assist with the transition. Despite all my struggles, I am strongly against government intervention in the development of sustainable agriculture. Germany is a case study in my reasoning. Germany subsidized the switch to encourage more organic production. Initially, many people switched, but for the wrong reasons. They switched to get the subsidy, not because they believed in the organic system. Once the subsidy ended, many switched back. If the government leads the transition, they will invariably focus on the supply side of the market; supply will lead demand, and farmers will switch to capture the incentive (farm the mailbox). Once the incentive stops, however, farmers will see an oversupply and switch to something more profitable. If the public leads the transition, demand will lead supply. Economic incentive will stay in the marketplace and gradually diminish over time as producers improve their methods and efficiency. Plus, part of the mess we're in is directly attributable to governments across the globe interfering in the agriculture and food system. Why should we expect government bureaucrats or their political masters to suddenly wake up and start getting it right?

What if the government imposed regulations rather than providing subsidies? You've probably yelled out "There should be a law!" at least once while reading this book, or maybe just thought it. I'm not sure how to break this to you gently, but there are already thousands of laws, hundreds of thousands of laws even. I used to be able to deadpan the line "I'm from the government, and I'm here to help" and everyone would get my meaning and chuckle. I can't say it anymore because everyone misses the irony.

The government is not the solution to any of these problems. I know where there is a composting facility that receives tonnes of perfectly edible produce almost every day. They're the culls from a nearby vegetable packing operation. Most of the onions received are simply too small to make the grade. Most of the carrots are too large to make the grade. This is government-enforced waste. All produce sold in Ontario must be graded prior to sale. It's a lose-lose situation. Taxpayer dollars are spent on inspectors, and perfectly edible food is discarded. The standards are there so that when a megastore orders a tractor-trailer load of Canada Fancy apples, they know precisely what they are buying. In short, grading primarily serves the industrial food system. Can you imagine the chaos that existed prior to grading standards? I can imagine people standing around in market squares trying to decide whether the carrots were large or medium. Actually, no I can't. I have faith that consumers can serve as their own grading inspector. If it doesn't look right, don't buy it. This is another paradigm shift we need. Consumers can look at most pieces of food and decide for themselves whether it is edible.

I'm not sure where this next line comes from (Joel Salatin uses it,[72] and so does Bill Mollison,[73] but I can't find an attributable source):

72. Joel Salatin has become one of the "poster boys" for sustainable food production in the US. He was featured in The Omnivore's Dilemma by Michael Pollan and the movie Food Inc. I first heard Joel speak at an organic conference over 15 years ago. To misquote a country song, I knew Salatin when Salatin wasn't cool. I've since heard him speak several more times, participated in workshops led by Joel, and visited his Polyface Farm in Virginia. I consider him one of the mentors for the Stoddart Family Farm.
73. Bill Mollison is an Australian who is considered to be the father of permaculture. He literally wrote the book on it in 1978: Permaculture One. He coined the term "permaculture" from a contraction of permanent agriculture.

"The most culturally subversive thing you can do is plant a garden." The next best thing is to educate yourself and form a relationship with the farmer that grows your food. By building a relationship, you're creating mutual accountability. The farmer will no longer be selling to a faceless corporation that will pulverize what he grew and reconstitute it as some form of heat-and-serve calories that tastes vaguely similar to something your grandmother used to cook. But understand that a little information is just dangerous; it takes a lot of information to make an informed decision. I've had people refuse to buy my chicken because the chickens weren't 100% grass-fed. I've had someone else refuse to buy my chicken because I couldn't guarantee that they had been fed a vegetarian diet. (Have you ever seen two chickens fighting over a cricket? You best not get in the middle; they're not averse to having human flesh on the menu.)

I can usually tell in the first five minutes whether a customer is going to be with us for the long haul or they're just trying something out because they heard it at their kaffeeklatsch. It's in the level of research that they've done and the questions they ask. I like informed customers. They understand why I can't sell them a boneless chicken breast without the bone and wing attached and why half the time the chicken quarter in their CSA share is a thigh and leg instead of a breast and wing. They like the fact that my chicken doesn't taste like the stuff in the store. Even some of these informed customers, however, are gobsmacked when they first see a 3.5-kilogram pasture-raised chicken.

We have some customers that have been with us since the very first day we sold at a brand new farmers' market in Toronto. They love our ethos. They love the fact that my ten year old can count back change the proper way without having a cash register tell her how much. They love opening a carton of our eggs and being surprised by the artistic creation that is a dozen eggs produced by heritage breeds of chicken. They certainly don't look like they were stamped out in a factory.[74] I love the fact that they're committed to buying from me, regardless of what happens to the price of pork bellies in Chicago. I love the fact that when Silvia is apportioning the shares each month, she can visualize many of the families that are behind the names on the list. These relationships create accountability in a way that laws and regulations can't.

---

74. In case one of the government inspector types is reading this. We don't sell chicken eggs at our farmers' market booth. You don't need to inspect us, again. I also trust that you have reviewed the requisite regulations and confirmed that it is perfectly legal to sell ungraded duck, guinea fowl, and quail eggs at a farmers' market.

Additionally, laws are written by people, and we have a minimal ability to accurately process the relative risk of infrequent events. A plane crash halfway around the world will headline the news and make some people nervous when boarding a plane. An automobile crash generally has to be spectacular to headline the news. Yet, you are far more likely to die in a car crash than a plane crash.[75] In Canada, we had a very illuminating example occur a few years back. A major meat company had a bad run of listeria-contaminated cold cuts. The outbreak headlined the news from July to December. There was video of people emptying all the meat (not just the cold cuts) from their freezers into the trunk of their cars and hauling it to the garbage transfer station. The outbreak officially claimed 23 lives, most of them people who were already in a compromised state of health. In the same span of time, over a thousand people in this country were killed in car accidents. We were more scared of the tainted meat than of driving our cars. The focus of the public outcry was on how a meat company could be allowed to ship contaminated meat (although no meat is free of bacteria). Very little focus was put on the nursing homes that were breaking protocol by serving cold cuts to people with compromised health.

We also find it difficult to regulate our actions when the action and the adverse outcome aren't directly connected in time. Most of us don't drive double the speed limit on the freeway because we know it will shorten our life expectancy, and yet most of us drink a soda every day even though we know it will shorten our life expectancy — but we won't go out in a flaming ball of fire; we'll gradually suffer any number of the joy-killing effects of diabetes.

Our public institutions have similar myopic views. Hot dogs and French fries are some of the easiest street-vending foods to get public health approval. One of the hardest is a salad. Why? A deep fryer is the holy grail of food purification machines. You can drop a festering piece of road kill[76] in a deep fryer, and as long as it is heated completely

---

75. Yes, for you statisticians out there, I know that if you rephrase the analysis as risk of dying per kilometre travelled, the risks are a lot closer to equivalent. According to a statistician, the average person in this country has slightly more than one breast and slightly less than one testicle. I've yet to encounter a statistically average person.

76. I probably grossed a few of you out with that statement. If that statement grossed you out, I wouldn't ask for the ingredient list on most of the stuff that you are eating out of a deep fryer or fast food kitchen. On the back concessions, hot dogs are often referred to as "pucker sausage" because both puckers are in there. (Only one actually is, but it makes a good story.)

through, you can safely consume it. (I'd probably peel the pelt off before you put it in the fryer — the pelt will be worth more if it hasn't been deep fried.) Unfortunately, there is nothing you can do to a salad and have it remain recognizable as a salad once it's been contaminated.

Public health officials are required to regularly inspect and approve the kitchens in our school cafeterias. They focus on cleanliness and freedom from obvious signs of rat infestation. They never comment on the content of the meals being served to our children. (We can't kill them quick, but we can kill them slow.) I've watched with amusement as the battle has spread across North America over what beverage should be in the vending machines in schools. Soda Inc. corporate types must find the venom and rhetoric amusing. At the end of the day, they'll sell soda, sweet fruit juices, or even bottled water in a vending machine. If you're in an argument about which sweetened beverage should be in the vending machine at your children's school, somebody asked the wrong question. I know that I went to school a long time ago (when we had to walk barefoot uphill both ways in the snow), but we had these marvellous devices that dispensed icy, refreshing water. Public health prefers a cold, sterile bottle of soda to the unhygienic, bacteria-covered water fountains. But technology has advanced; you don't even have to touch the most modern fountains. Why should we force people to drink soda because it's cheaper than bottled water when every building in every city is hooked up to a potable water supply? The boys and girls over at Soda Inc. must have split a gut laughing when they realized that they could put the water in a bottle, leave out the syrup, the colour, the flavourings, and the carbonation and then raise their prices. It is true that they have some expenses filtering the chlorine out of that city water they're bottling. I wonder why the labels all have pictures of glaciers rather than the intake pipes for the city water plant. Chicago's water intakes in Lake Michigan are quite picturesque at sunset.

I'm not convinced that many at public health know the difference between something that is going to shorten our life expectancy and something that won't. Certainly the rules they enforce have a certain bureaucratic inconsistency to them. We travelled to Europe a few years ago. I went to the supermarket while we were staying with friends. (I find most European supermarkets more interesting than most of the tourist hotspots.) They asked me to grab a dozen eggs while they picked up some vegetables. Ten minutes later, I met them at the checkout with empty hands. I had searched every cooler twice and couldn't find any eggs. They laughed. Eggs aren't kept in refrigerators. They're in the dry goods aisle. To put a further exclamation point on my point, if I'm operating a federally inspected egg-grading station, I have to store eggs at 10°C or below. As soon as those eggs enter a retail space, which are inspected by public health rather than the feds, the eggs have to be stored under 4°C. I hope nobody tells public health about the nest boxes my chickens lay in. I don't think I could convince a hen to consistently lay anywhere that is only 4°C (unless it's the warmest place she can find).

Our obsession with sanitary conditions might, in a rather ironic twist, end up killing us. Our bodies are very efficient bacteria-killing machines. We have defence mechanisms if we inhale, swallow, or stab bacteria into our bodies. However, those defence mechanisms need to be calibrated, preferably at a young age. There is now conclusive evidence that children raised on animal farms have a lower incidence of asthma (an overreaction of an improperly calibrated defence mechanism). Why? Farm kids put shit in their mouths. (Literally. OK, I think the researchers termed it "dirt" rather than shit, but on an animal farm there is a very fine distinction between the two. I suppose there is a stage called compost in between.) But are headlines screaming that parents everywhere should toss out their sterilizing wet wipes and alcohol sanitizers and let their children get down and crawl around on something that hasn't been drenched with Lysol? No. Instead, researchers announced they were working to develop a vaccine that would stimulate the same parts of the immune system as getting dirty when you're young does. They were looking for something they could sell you, not the most effective solution.[77]

Democracies are interesting beasts. The majority wins, but there needs to be a counterbalance to ensure the majority isn't tyrannical in its treatment of the minorities. I don't think that the alternative food movement — whether you define that as local, organic, non-GMO, humane, non-industrial, pesticide free, or grass-fed — has much chance of legislating many meaningful changes to the food system. The primary

---

77. If you want to delve deeper into the concept that farm kids playing in shit paves the path to world domination, I suggest you read *Guns, Germs and Steel* by Jared Diamond. According to his thesis, Europeans conquered most of the known world because they liked to cuddle up with the sheep and the cows in the wintertime, which built their immunity to a number of zoonotic diseases. They were then able to defeat other civilizations using biological weapons, such as smallpox. He is a little wordier, but the point is the same.

benefit of their lobbying and information campaigns is educating more consumers. When you argue for legislative changes, you have an opponent with vested interests in the status quo who will try to either quash your legislation or water it down to the point where it will not reach your intended goal. Democratic politics can be a nasty beast. Otto von Bismarck is quoted as saying, "Sausages are like law. It is better not to see them being made." I've seen both. I've participated in both. I'll take sausage making any day.

Governments in North America tend to magnify majority opinion. It's been over 30 years in Canada since the prime minister had the support of a plurality of those who voted. The US recently had a president that the majority of the citizens voted against. When only 2% of agriculture is organic, those views rarely get more than a polite inclusion in public policy formation. An odd bone gets tossed our way, but the majority of agricultural policy is formed to meet the interests of the major agricultural commodities.

If we could successfully legislate, what would we legislate? Even though I've hopefully made some convincing arguments supporting my points of view, I wouldn't want them enshrined in law. I'm certainly glad that my views 15 years ago weren't enshrined in law. I expect to learn as much in the next 15 years as I have in the last 15. In Holistic Management, the first step after implementing a plan is to assume that you're wrong and monitor for discrepancies between reality and the plan. You don't hear a politician stand up in the House and say, "We've been monitoring our new legislation and can see where we made some mistakes. We knew we would, we just didn't know what they would look like."

There's another tool that runs in parallel with legislation. It is the marketplace. While we can decry the influence of multinational corporations on the goods that are offered and the policies of government, we can also use the very tools of those imperial capitalists to subvert their purpose. We have a vote with our dollars every day. When it comes to food, we have a vote three times a day. Nobody can cancel your vote, and your vote isn't meaningless if your choice doesn't get the largest market share. Nobody forces you to eat what the highest market share producer produces from now until the next election. Your vote for alternatives will serve to reinforce the availability of those alternatives. If you're able, willingly pay a premium for the alternatives that you believe to be the healthiest, most sustainable choice knowing that your willingness to pay the premium will eventually attract new producers and that the premium will diminish. That's the way the capitalist system works. We can't legislate what we want, but we sure as hell can starve the enemy to death. Withholding dollars from corporations is equivalent to withholding oxygen from people.

Some are of the opinion that, because each household spends only a couple thousand dollars a year on food, their consumer decisions can't be meaningful. Every spring a river or rivers overflow their banks. Have you ever noticed what people use to protect their homes from the millions of litres of water that are about to flow across their property? A sand bag. Each sandbag weighs around 25 kilograms, and when you stack enough of them, you can hold back the pressure from millions upon millions of litres of water weighing thousands of tonnes. How do those sandbag dikes get started? By filling the first bag with sand, then another one and another one. Think of your grocery bag as one of those sand bags. Every time you fill your grocery bag with sustainably produced real food, you are throwing one more bag on the dike to hold back the destruction that will be wrought if the river of unsustainable food production practices is allowed to flow unchecked. You will be assisting in the conversion to a truly sustainable food system.

CHAPTER 13

DOING THE RIGHT THING

"Efficiency is doing things right; effectiveness is doing the right things."

— *Peter Drucker*

I spent seven years working for management consulting organizations. One of my specialties was process re-engineering — making processes more efficient. I had all kinds of fancy tools: time and motion studies, process mapping, value-added analysis, etc. The first thing I always had to do was make sure a company wasn't focused on doing things right but rather focused on doing the right things. Put another way, it doesn't matter how efficient a process is if the process isn't delivering what is needed. It's an easy trap to fall into.

The classic example that is often cited in training programs is the case of a large computer installation and the requirement to generate a particularly troublesome report. After spending considerable time trying to figure out how to make the new computer system generate the report, the project manager decides to ask why the report is needed. The analyst says she needs it to report to her manager. The manager says he needs it to report to his VP. When the VP is asked, she says, "I don't know why they keep sending me that report each month. I never read it."

Because agriculture is largely a commodity business, the key to long-term profitability has been lowering the cost of production either through increasing yields or reducing costs. We have become very efficient at producing the main commodities. If you look at the history of feed conversion in the pork industry, it has been reduced from 3.5 to 2.5 in the last 60 years. On the surface, this change in conversion sounds like a great thing. It takes one less kilogram of feed to grow each kilogram of pork than it did 60 years ago. If you then combine that feed savings with the increase in yields of corn and soybeans over the same period, you can calculate the number of acres that don't have to produce crops to feed the current swine herd. Through the lens of doing things right, we have become a lot more efficient.

The pig's original role on farms, however, was to turn waste into meat and fertilizer. In 1950, we weren't feeding pigs with 100% field crops. Pigs ate the waste out of the kitchen, the shorts when wheat was cleaned for seed, and the whey separated from the cream. They rooted around the orchards cleaning up windfalls, and they got a little grain to top it up. Wheat was only fed to pigs if it was "feed grade," meaning that for one reason or another it was not good for anything other than feeding it to pigs.[78] Even as agriculture began to specialize in the 1960s, a number of entrepreneurial hog farmers in the shadows of urban areas were utilizing pigs' ability to recycle food waste into pork and fertilizer. These farmers gathered the food waste from restaurants and bakeries and then fed it to their pigs. Today, companies and governments are spending millions trying to develop methods for recycling "wet" waste. We already have pigs, and we use to have a process.

This process vanished as the pork industry worked hard to have garbage feeding banned because feeding pigs "garbage" tarnished the image of the industry. There was some concern of disease being recycled back through the pigs, but forcing the farmers to cook all the waste before feeding it alleviated this problem. Instead of investing in huge fossil fuel sucking composting facilities, we should be diverting that waste stream to existing hog farm infrastructure for direct recycling into food. But that would be doing the right thing rather than doing things right. Instead we invest millions each year in fine-tuning the amino acid and energy profile of hog feed to increase efficiency.

The idea of feeding grain initially to pigs and chickens and then to cattle and sheep is a relatively new one. Prior to cheap fossilized solar energy (oil) and governments interfering in agricultural markets, grain was relatively much more expensive and wasn't wasted on being fed to animals in great quantities. To give you some perspective on how cheap oil and the advent of tractors impacted agriculture, I'll relate a comment

---

78. The most common reason for wheat grading "feed" is the kernel has started to sprout in the head before harvest. Once the kernel starts to germinate, enzymes are activated that degrade the gluten and significantly reduce the baking quality of the wheat.

my father made when he first started working for me. The first tractor I bought after I took over was a 280-horsepower, four-wheel-drive Versatile. It was old — made in 1978 — but in terms of straight-ahead horsepower for dollar spent, it couldn't be touched. I put a 12-metre cultivator behind it, and we could cover some ground in a day.

The first day dad drove the Versatile, he asked me, "Am I right in thinking I'm covering a little over 20 acres an hour?"

I said, "Yes, why?"

He said, "I couldn't do that in a long day with a team of horses."

There was a reason fields used to be ten acres in size.

Old MacDonald's farm was a very efficient enterprise, but in a completely different way than agricultural efficiency is defined today. Synergies between enterprises were maximized. The efficiency of the whole was maximized rather than each of the individual enterprises. The largest percentage of a farm was planted to hay and pasture. They converted solar energy into biomass. That biomass was turned into human food by cattle, human clothing by sheep, and work energy by horses. Some grain was grown — wheat for milling and oats to top up the horse feed.[79]

Another example around pigs is their inefficiency with phosphorus. When we had small farms with a few pigs converting waste into meat and fertilizer, the phosphorus in pig manure was considered a good thing. The pigs were excellent at recycling phosphorus back to a plant-available form. When we started focusing on making pigs grow faster and convert feed more efficiently, we found out that gains could be made by adding extra phosphorus to their diet. However, when you have thousands of pigs in one spot and are hauling in their feed as well as hauling in extra phosphorus to feed them, the phosphorus in their manure becomes a liability. Science's answer was to genetically engineer the "Enviropig." This pig uses phosphorus much more efficiently, and the level of phosphorus in their manure is reduced by 50%. In restructuring the paradigm, we turned a positive into a negative and then created a potentially risky solution to a problem that doesn't have to exist.

---

79. Horses are not ruminants. Ruminants have pre-digestion fermentation in their rumen and breakdown and absorb nutrients through four stomachs. Horses have post-digestion fermentation. They have a cecum between their stomach and intestines that works to break down some of the cellulosic material so that it can be absorbed and utilized. However, it is a much less efficient process than the ruminants.

The Western mindset is blinding in its pursuit of efficiency, where each new problem has a distinct solution. I used "blinding" advisedly in that sentence — "brain numbing" might have been a better phrase. How else do you explain an agricultural paradigm that hauls corn by the trainload from the Corn Belt to the Carolinas to feed pigs in barns with cheap labour, where the fertilizer (manure) flows out into lagoons and magically "evaporates," and then the pork is hauled back to cities in the Corn Belt? How do you explain growing vegetables in the middle of a desert to exploit illegal labour while diverting entire river systems and draining aquifers to irrigate the land just so the vegetables can then be shipped by refrigerated truck out of the desert, across the mountains, across another desert, and to the cities for people to eat them? We also put 15,000 dairy cows in a feedlot in the middle of a desert and haul feed to them so we can haul the milk across good pasture back to the cities. Do you know how much water a dairy cow drinks in a day? A lot.

Yet, from a sustainability point of view, the biggest problem with hauling corn by the trainload to hog farms and then hauling meat to the cities is not the fossil fuels consumed. Instead, it is the nutrients that are taken. Each kernel of corn is composed of carbon, oxygen, nitrogen, phosphorus, potassium, and many other micronutrients. A good portion of them comes out the back end of the pig. The phosphorus is precious and yet destructive. We've created two problems: First, the phosphorus is a problem in the area of the hog farms because there's too much of it and it becomes a pollutant in the waterways. Second, the farms in the Corn Belt have to replace all those nutrients they loaded on the train and sent to the hog farms. For the past 50 years, that's been cheap to do. Phosphorus supplies were abundant, and the energy to extract, transform, and transport it was cheap. Before this, manure was never a problem; it was valuable right where it was produced.

The next example is almost too fantastical to believe, except that one of my hired men had previously been a long-haul trucker and did this run a few times. He had a run from Ontario to Florida and back. His cargo was tonnes of carrots — both ways. Yup, he was hauling carrots from Ontario to Florida and then hauling carrots from Florida back to Ontario, usually in the same reefer trailer both ways. Hundreds of gallons of diesel fuel were burned each time, and the same tonnage of carrots was hauled to both places. The purpose of his runs was to haul carrots out of storage from Ontario to Florida and then take green tops from Florida to Ontario. Carrots out of storage are sweeter and therefore in demand in Florida. Green top carrots look fresher and therefore are in demand in Ontario in the winter.

Was society better off? My economist friends would say it was because there was enough profit in both transactions to pay the fuel bill and have some left over — it's called market arbitrage. That's why I tend not to identify myself as an economist much anymore. I can't think of anyone else who would say starting with 40 tonnes of carrots in Ontario and driving a truck 5000 kilometres to end up with 40 tonnes of different carrots in Ontario made anyone better off (unless you have stock in an oil company). The good thing about rising fuel prices is that local markets are less open to competition from afar simply because the cost of transportation goes up. The kind of idiocy in my examples gets reduced.

What does the "right thing" look like when it comes to agriculture and food production? I turn back to Allan Savory, not for his thinking on planned grazing but for his thinking on Holistic Management. I'm also a firm believer in defining "the right thing" as a set of principles rather than rules. Before we start talking about agriculture and food, we need to define our goal as a society — in Holistic Management parlance, the holistic goal.

A holistic goal has three components: quality of life, behaviours and systems, and future vision. This is the starting point whether you are talking about a farm, a person, or a country. Defining the quality of life you want essentially itemizes what you value. The behaviours and systems are what must be created to provide the quality of life that is desired. For example, if my quality of life includes financial security and strong relationships then I must produce profit (income in excess of expenses)[80] and time available to build relationships. The third component is a vision of what the distant future will need to look like from three standpoints: behaviour, the community, and the resource base. For example, in the future I want to be known as a just and honourable person; I want to live in a peaceful community with access to good education, healthcare, and economic opportunity for its citizens; and I want the land I manage to be improved and the wealth of our family to have increased.

---

80. This definition holds whether we are talking about a business, a family with a wage earner, or a government.

If we look at how society might define a holistic goal for the agriculture and food system, I would contend that at a high level it would look something like this:

Quality of Life	Behaviours and Systems	Future Vision
Everyone has access to an adequate supply of calories, protein, and nutrients.	Sufficient calories, protein, and nutrients to meet everyone's needs must be produced. Distribution of the calories, protein, and nutrients has to ensure that everyone can meet their needs.	Behaviour: The food system is seen as serving the needs of humanity in terms of both provision of food and caretaker of the resources used to produce the food.

Future resource base: The environment must be able to provide all of the ecosystem services required for all living things on the planet to flourish.

Community: The agriculture and food system is part of a global community that provides equal opportunities for health, education, and security regardless of geography or occupation. |
Everyone has an assurance that their food is safe.	A functioning and trusted system to ensure the safety of the food supply is protected.	
An adequate standard of living exists for everyone to meet his/her basic human needs.	All workers in the food system must earn an adequate standard of living.	
There is security of person and property.	The food system must be seen to be just, and individuals must have access to an impartial and trusted system to solve disagreements and disputes.	

We then work forward to define how the food system needs to operate to achieve our holistic goal. The first thing that should be immediately obvious is that our current system is not producing any of the elements of the quality of life we desire for the citizens of the world. I would also argue that we are currently moving away from the future vision rather than towards it. Agriculture's treatment of resources is better characterized as exploitive than as a caretaker. Agriculture is contributing to the loss of habitat, endangerment of species, increasing atmospheric carbon dioxide levels, and loss of soil globally. Our current trajectory does not lead to a flourishing future. In both the poorest and wealthiest nations of the world, people in the food production system are over represented among a nation's poor. Agriculture is the key human development that facilitated the creation of the wealth and standard of living we currently enjoy, but those involved receive the least share of the prosperity.

I've mentioned another feature of Holistic Management that I think is key: Plan, but assume that you are wrong and monitor for the first signs that a course correction is needed. I have ideas about what an agricultural system looks like that will meet the holistic goal, but I doubt that I have all the components right for even one geography. What follows is where I would start the plan and then monitor for failure.

First, we have to stop wasting soil. Soil degradation takes the longest to heal of all the damage that is being done. The most obvious change is to stop tillage. This will be a two-step process. In the short-term, no-till or zero-till can drastically reduce the soil degradation caused by our current crop mix. Long term, our food crop mix needs to shift to perennial crops. You can influence this shift by changing your diet to include foodstuffs derived from perennial crops (fruits, nuts, perennial vegetables) and meat that has been raised on perennial crops (grass-fed beef, lamb, and chevon, and pigs and chickens primarily fed on waste streams from perennial crops).

Second, we have to stop depleting our water resources. Agriculture consumes 70% of the annual take of water globally. Agriculture needs to return to primarily rain-fed methods other than in regions where the irrigation water can be supplied as part of the annual rainwater endowment. Agriculture also needs to work to maintain or restore the hydrological cycle in arid and brittle environments. The good news is that the recommendations for conserving water are aligned with the recommendations for conserving soil. The best way to maximize the capture and use of rainwater is to establish permanent-living vegetation on the land.

Third, we need to stop polluting both our land and water resources. Maximizing land-based production of crops is not a valid reason for destroying water habitat. Three-quarters of the Earth's surface is water, and our ability to harvest food from that water has greatly diminished in the last century. Agriculture has contributed to the decline. There is a developing body of research that clearly shows the claimed benefits from pesticide and synthetic fertilizers in agricultural production are greatly exaggerated and possibly nonexistent. When you factor in the externalities created by their use, they are inconsistent with a flourishing future for all living things. Organic is the primary alternative currently available to reduce the pesticide and synthetic fertilizer footprint of your food. However, conservation of soil and water do not automatically flow from an organic choice. You will need to understand the production methods used on the farm before you can make a fully informed choice.

Fourth, we need to understand that there is no such thing as "waste." Everything that is discarded simply becomes an input to another process. Agriculture is an industry that can consume all of its waste; however, it is currently a long way from that ideal. We need to get past our squeamish superstitions and embrace closing the loop on food production. Vegetarian pigs and chickens are a luxury the planet cannot afford. We must find a way to recycle all the nutrients in sewage. "Food garbage" is a valid source of feed for pigs and chickens. Support solutions that close the loop.

Fifth, we need to switch from an industrial paradigm to an ecosystem paradigm when we think of agriculture and food systems. Complexity, distributed design, and diversity are all contributors to resiliency in ecosystems. Every ecosystem can largely function on the local water cycle, local mineral cycle, local population dynamics, and local energy flow. While all the ecosystems are integral parts of the whole, the whole can withstand the loss of any local system and regenerate it. There are actors that traverse the boundaries of local systems, but the core functions are supplied locally. To close the loop on our agriculture and food system, we must maximize the amount of recycling that can occur in place. The closer we are to our food, the closer our food is to its food. This is not to say that international transport of all food should be curtailed, but when that transport represents a material movement of nutrients, it has to be matched by a return trip for the nutrients. We need to use the ecosystem paradigm at all scales of production from the farm to the community to the international movement and consumption of food. Monoculture crop production is the antithesis of an ecosystem paradigm. Almost everything on store shelves today is connected to monoculture production — from corn, rice, and wheat to tomatoes, peaches, and celery to beef and pork. There is no well-functioning ecosystem that is dominated by a single organism. The global agriculture and food system is gradually being dominated by a small number of giant "organisms" (companies). To reinforce an ecosystem paradigm, you need to choose food from sources that are diverse and that understand they are a part of the local ecosystem. As well, you need to reinforce the diversity of your local "food shed" whenever possible.

Sixth, we need to rethink the value we place on food and the people who produce it. The first people that have to be sustained by the food system are the people who work in it. A farmer will only invest in preserving the land for future generations if he has sufficient profits to make that choice. If you care about future generations, you should be willing to invest in the food production systems that they will thank us for.

Seventh, we need to understand that agriculture and food are inherently powered by renewable energy. Photosynthesis transforms solar energy into carbohydrates. Our reserves of carbon-based fuels are being wasted when applied to agriculture. We know we have finite supplies of carbon fuels but an infinite demand for food over time. The long-run solution does not include carbon-based fuels, and therefore, the short-run solution should be working towards eliminating the use of carbon-based fuels. The best place to start is to eliminate processed foods from your diet. Localizing your diet helps, but it doesn't have as large an impact as some would have you believe. (Localizing your diet is good for many reasons, but I wouldn't count reducing the energy footprint of your diet among them.) Again, we see some overlap with other dimensions. Tillage is a waste of fuel. No-till methods are working for most crops if there is a commitment to the method. Synthetic fertilizers are a waste of fuel. Penning ruminants in a confined feeding operation and hauling them feed is a waste of fuel. Annual crop agriculture is a waste of fuel. The only label in a store that will give you an indication of the fuel consumption of your food is a processed food label — they all have more embodied energy than their raw ingredients. After that, you need to understand the farm that is producing the food.

Finally, agriculture is part of the solution. Our world is facing a number of potential crises. Agriculture and food production are at the heart of the solutions to many of them. Agriculture can reduce atmospheric carbon dioxide levels. Agriculture can maintain or restore the hydrological cycle in brittle environments. Changes to agriculture can improve water quality and improve the production of water-based ecosystems. Improved agricultural production can alleviate poverty and hunger globally. More abundant, well-distributed food, reduced poverty, cleaner water, and reduced climate disruption all translate to reduced conflict. Your choices will push agriculture towards being the solution or the problem. But we need to focus on the right things.

CHAPTER 14

LOOK! SOMETHING SHINY!

The power of population is indefinitely greater than the power in the earth to produce subsistence for man.

— *Thomas Malthus*

Kasne has made it his personal mission in life to ensure that none of our cats succumb to the feline diabetes epidemic that is sweeping North America. He keeps them all very fit and in top-notch cardio condition. I don't think he's malicious; a border collie is just so full of energy that it will chase anything that will run away. Cats always run away. In the case of one cat, however, Kasne's intentions seemed a little more malevolent. The cat's name was Moses, but I called him our "mafia don." He sat up on the beams in the old bank barn striking a pose that was reminiscent of a cross between Marlon Brando and a Cheshire cat. He clearly was well fed, and he generally managed to avoid participating in much exercise.

One day, I entered the bottom of the barn with Kasne close at my heels. Kasne darted past me. A cloud of dust formed amid much growling, hissing, and caterwauling. When I got closer, Moses was halfway up one of the support posts, spread-eagled with all four claws stretching up and gripping the wood as if his life depended on it. It did. Kasne was stretched almost his full length, with Moses's tail firmly gripped in his teeth. I learned a few cuss words in cat dialect in that moment — or maybe they were death threats. Kasne momentarily forgot what he was doing and decided to bark at Moses. In doing so, he released his grip on Moses's tail. It was as if Kasne had released an arrow from a bow. Moses shot straight up the post, bumped his head on the ceiling, and just managed to recover enough to catch himself on a cross beam, safely out of Kasne's reach.

If Moses was still around, he'd likely put a hit out on me if he knew I was about to tell you this next story. Moses lived in the house when he was a young kitten, along with a rabbit named Annabelle. But Annabelle was older and therefore larger than Moses. She would chase Moses around the house. I'm not sure if the scars from those experiences ever healed. Moses always seemed to have an unhealthy interest in the kids' pet rabbits when he got older. Annabelle chased Moses because he would run. I think she thought it was a game, though Moses generally looked like he was running for his life. Kasne chases cats because they run. The sheep will chase anyone with a pail because they might get some grain from it. It doesn't matter if you're leading them into the barn to be penned up for shearing; the sheep focus on the promise of grain in the bucket.

As we traverse the long and winding road to a sustainable food system, many things will try to capture our attention to pull us off the path. There will be many groups trying to divert our attention from the goal of true sustainability. They will distract us with false concerns and false solutions. I already explained why I think GMOs are on this list for food production. In this chapter, I discuss two more shiny objects.

FEEDING THE WORLD

If you believe Big Ag, we need synthetic fertilizers, pesticides, and biotechnology to keep people from starving. Choosing organic and sustainable agriculture is sentencing people to starve to death! Can you say red herring? Malthus first postulated this theory over 200 years ago.[81] It's resurfaced many times, and we've never made it there yet. Increases in food production have outpaced increases in population resulting in increasing per capita production of food.

There's also a logical problem with the argument. Since conventional agriculture is more dependent on nonrenewable resources, it will exhaust those resources sooner and the world will run out of food sooner than the alternatives. If we accept conventional agriculture's claim that it produces higher yields more efficiently, whether we are better off with their

---

81. "The power of population is so superior to the power of the earth to produce subsistence for man, that premature death must in some shape or other visit the human race," (Thomas Malthus, "Principle of Population").

methods depends on your time horizon. In the short term, there is more food for more people. In the long run, however, there is substantially less. Furthermore, the externalities created by conventional agriculture — phosphorus loading of waterways, organic matter consumption, erosion, and greenhouse gas emissions — leave the environment less able to produce food. The longer conventional agriculture continues, the faster we exhaust nonrenewable resources and the greater the damage done to our ability to use sustainable methods to produce food in the future.

There's another angle to the Malthus argument resurfacing at this point in history. Call me cynical, but the arguments have resurfaced in earnest at exactly the same time we have an agricultural commodities bubble growing. A lot of capital has moved into agricultural commodities and agricultural input commodities in the last several years. The Malthus argument is an excellent story to explain why food is a good investment. This is the third run up in agricultural commodities that I've witnessed since I started farming. Both times previously, the "experts" said we had achieved a new pricing plateau. They were wrong.

The start of this bubble was fuelled by the ethanol subsidies driving corn demand up. It's continued because speculators have made some nice coin on the run-up. The primary players in commodities markets used to be the producers, traders, and processors of the commodities. Wall Street has now created investment vehicles in commodities that look a lot like stocks. Futures markets bear little resemblance to stocks. Every "investment" has a short timeline and has to be settled with the winners and losers declared. There is no "buy and hold" strategy for futures contracts.

Goldman Sachs created the first commodity futures index with properties that look a lot like an investment in stocks. In 2003, there was a paltry $13 billion of contracts in commodity futures. In the first 55 days of 2008, as subprime mortgages were melting down, $55 billion poured in and by July 2008 $318 billion were "in commodities." This $318 billion was doing nothing other than betting that commodity prices would go up. There weren't any more acres under production. It didn't represent additional money invested in food production. It was just bets being placed on the outcome of commodity prices. The problem is that they have to keep drawing new money to the party to be able to roll their positions forward. Bernie Madoff knows a lot about this game. The only difference is that the big investment banks have laws backing them that say their particular kind of Ponzi scheme is legal. There was a tech bubble. There was a subprime bubble. There is a commodities bubble. It will burst. Sorry, I'm an economist, so I can tell you what will happen or when it will happen but not both.

There are those who create wealth and those who live off the creation of wealth. We have too many of the latter and too few of the former. Farmers aren't bidding the price of food up, and we're not really gaining much from the bubble —our inputs have been bid up to historic highs as well. Those that have created this bubble are bunch of hired goons sitting in financial towers, who wouldn't know the difference between a pile of compost and a load of shit. They'll get what's coming to them. The only question is whether we'll let them drown this time. All the "wealth" sitting in commodity futures contracts is just digits in brokerage accounts. They can go down faster than they went up. When new money stops flowing in, the scheme will be discovered. The capital will fly out of the market faster than a scared duck. There will be much stamping of feet and gnashing of teeth, and we'll probably convene government inquiries. The thieves will simply walk away and start to inflate another bubble.

As an economist, I graduated from university a staunch supporter of laissez-faire economics. The longer I've been an entrepreneur, the more I've seen imperfections in markets that have occurred despite and because of government intervention. We're not the first era to have financial "bubbles" founded on nothing but a good story and public gullibility. Unfortunately, most government interventions in markets are politically motivated and rarely focus completely on the public good. Agricultural commodities are probably the market with the most government intervention and yet they are still imperfect, largely because government interventions have been the result of industry players asking for interventions that benefit them rather than the public good.

If governments and markets could solve the issue of feeding the world's population, they already would have. There were massive grain surpluses in North America and Europe through the last half of the twentieth century. It is true that the population cannot continue to expand forever. If it did, at some point humans' demand for food would exceed the earth's ability to capture energy and convert it into edible calories. Many fiction writers have hypothesized how future humanity might manage the challenge of keeping the population from expanding beyond a certain level. Eliminating poverty is the easiest way to stop population growth. The birth rate in most developed countries is below the replacement rate. Without immigration, the populations of these countries would be declining. We see the same trend in less developed countries — as the income levels and life expectancies increase, birth rates decline. UN population estimates predict the global population levelling off within this century.

Proponents of high-input industrial agriculture are convinced that the solution to the problem of feeding the world is found in taking Western industrialized agriculture and applying its methods in all the regions of the world where agriculture is primarily low input or subsistence. We've gone down this road before. Norman Borlaug's "Green Revolution" was supposed to have solved the problem of feeding the developing world half a century ago. It didn't succeed because food insecurity often isn't a function of yield. The UN recently released a report on child malnourishment that showed higher rates of malnourishment in India than in sub-Saharan Africa. Opponents of transferring high-input agriculture to the developing world immediately seized upon the report as "proof" that the Green Revolution had failed because India had embraced the Green Revolution and sub-Saharan Africa had had minimal impact from it. I hope the people who drew those conclusions never stand too close to cliffs. When you looked deeper into the data, there was a significantly higher rate of malnourishment in females than in males in India. My conclusion would focus on the still firmly entrenched caste system in India and the imbalance in worth placed on males versus females.

So why then do I oppose taking high-yield cropping methods to the developing world? Annual crop agriculture is a single-hand game of poker where the farmer goes all in before he's been dealt a single card. To survive a decade, you have to win ten straight hands. Would you play a single hand of poker for your entire paycheque each year for the next decade in a game in which losing a hand means you lose your job? The analogy is a little hyperbolic in the context of North American agriculture. North America has fairly reliable weather patterns in most areas and the incidence of truly devastating weather is minor. Even in the stable temperate zone of North American agriculture, however, high-input cropping systems cannot exist without government policy, support programs, and a well-developed banking infrastructure. It's an easier bet to make if the government guarantees you at least a tie and there is someone on the sidelines willing to lend you the money to place the bet next year. The generosity of government support programs in North America and Europe is unmatched in the developing world. Similarly, the stability of North American banking systems and the availability of debt financing is seldom found in the developing world. High-input agriculture has a high external demand for cash to purchase the inputs.

In mathematical economics, there are two general theories of managing for long-term success: max–max and max–min. In the max–max strategy, the objective is to apply inputs at the level that will maximize your profit in your best year, knowing that you are over applying inputs in years where weather conditions are suboptimal for maximizing yield. The max–min strategy applies inputs at a level that will maximize your profit in your worst year. In other words, this strategy manages to avoid steep losses when weather conditions are unfavourable, knowing that you will be under applying inputs in years in which weather conditions are optimal. The max–min strategy is focused on surviving in order to plant another crop rather than winning the lottery with a bumper crop.

After the North American farm crisis of the early 1980s, there was a line of research[82] in the US that compared a low-input sustainable agriculture (LISA) model against the high-input/high-yield model. On average, the high-input/high-yield model was more profitable; however, when you simulated farm survival, LISA was the clear winner. Poor years occurred sufficiently frequently to bankrupt high-input/high-yield farms a much higher percentage of the time. In regions where the weather can be devastatingly erratic and generating surplus cash is challenging, I think local models akin to the LISA system should be adopted. The United Nations Special Rapporteur looked at agricultural systems in the developing world and came to a similar conclusion.[83] Smallholding, diversified, sustainable agriculture is capable of producing higher yields more consistently than the imposition of high-input/high-yield methods.

As an example, an aid worker living in Madagascar observed local Malagasy farming methods for a decade and gradually developed a system of rice production that he called the System of Rice Intensification (SRI). It involved spacing the rice plants farther apart, keeping the soil moist rather than having standing water, and applying compost from readily available organic waste. The results of the new method were decreased input costs (fewer rice plants, zero-cost fertilizer, no pesticides, and less water), improved yields (50% to 100% higher yields), and less work for the farmer (fewer plants to plant). The wider plant spacing also meant a simple, locally built tool could be pulled through the field for weed control

---

82. Clive A. Edwards, "Integrate Systems."
83. United Nations Special Rapporteur on the Right to Food, "Agroecology."

rather than repeated manual hoeing. SRI has now been tested in at least 28 rice-growing countries, producing similar results. The current world record for rice production per acre is held by a farmer using the SRI system. SRI is successfully being used to increase rice production in the aftermath of the earthquake in Haiti. Growers have also successfully applied the same principles to growing sugar cane and millet.

The developing world has a chance at a "leapfrog" development path in agricultural technology similar to how they leapfrogged wireline telecommunications and went straight to cellular telephone technology. In agricultural technology, they can learn from the mistakes of Western agriculture, leapfrog the environmental destruction wrought by high-input cropping systems, and go straight to sustainable systems. These systems will build their soil, increase their food security, and improve their economies.

ETHANOL AND BIOFUELS

There's an old joke from the era when junior mining stocks were the primary high-risk/high-reward stocks on the market: "What is the definition of a mine? A hole in the ground with a liar standing at the entrance." From my perspective the only thing that has changed with the passage of time is where the liar is standing. In the late 1990s it was Internet companies, then it became biotech and biofuels companies.

Biofuels are another "something shiny" that many people are trying to distract us with. The ethanol industry has come a long way in the last decade. They now have systems that are energy positive. The ethanol process at the turn of the century didn't create a surplus of energy — it was merely a method of turning diesel fuel and natural gas into "renewable" ethanol fuel.

The latest ethanol energy balance assessment by the US Department of Agriculture (USDA) states that for every BTU of energy input into ethanol production, we get 2.3 BTUs of energy output. We'll use this number even though the USDA is a large proponent of ethanol production and tends to make favourable assumptions in their calculations. I'm going to use it because even with this optimistic number you will clearly see how absurd it is to think that ethanol will replace any material percentage of our fossil fuel consumption.

With a 2.3 to 1 output ratio, we would need to dedicate 43% of our land base to ethanol production to produce enough energy to grow the crops for fuel, process the ethanol, and grow crops on the remaining 57% of the land base. Most estimates I've seen suggest that agriculture's consumption of energy as a total of all energy consumption is in the range of 2 to 3%. Approximately 87% of the world's energy consumption is from fossil fuels. Using the 3% energy consumption rate, to replace all fossil fuel energy sources with ethanol would require over 12 times the current land area in agricultural production. Even if we assume that ethanol energy efficiency gains an order of magnitude through efficiency improvements, we would need 1.2 times the existing agricultural land base just to produce enough energy. Despite the advertisements that the corn and ethanol industries are running, it is a choice between food and fuel. We can't do both, and they can't do math. To get to a point where it even becomes close to reasonable to consider ethanol as a viable renewable fuel, the energy balance would need to be over 200:1. Even then, we would require over 10% of our current agricultural land base to supply our energy needs, assuming they remain constant. Further evidence is seen in the current ethanol industry; 40% of the US corn crop is used to produce ethanol. The current level of ethanol production is insufficient to meet a 10% renewable fuel standard for gasoline.

Biodiesel is 50% more efficient, producing 3.2 units of energy output for every unit of energy input. So we would only have to dedicate 30% of our agricultural production area to produce enough biodiesel to grow crops on the other 70%. We would only need 8.7 times our current agricultural land area to produce enough biodiesel to replace all fossil fuels. Again, we would need an efficiency gain of two orders of magnitude before the land area dedicated to biodiesel production would drop below 10%.

The one source of biodiesel that does make sense is the conversion of waste vegetable oil to biodiesel. This converts a waste stream into a usable commodity; however, we don't need to go to the expense and environmental risk of converting the vegetable oil to biodiesel. There is a much simpler option: dual fuel diesels. This requires a fairly simple modification of diesel vehicles so that they have two tanks and a method of switching between them. The primary issue with using straight vegetable oil in diesel engines is that it is too viscous at room temperature to properly flow through injection systems. By switching to a dual fuel system, the engine is started on diesel fuel. The engine heat is then used to warm the vegetable oil, and the engine switches over to vegetable oil. At shutdown, the process is reversed so that all the lines are flushed with diesel. The engine is run on diesel for the last few seconds, leaving it ready to start on diesel. There is no lye, no ethanol, and no precise chemical reaction to manage. There is also no risk of damage to engine components from unreacted lye or ethanol and no risk of environmental damage from the production and transport of lye.

There are proponents of a number of magical bullets. Some argue we should turn to cellulosic ethanol, which uses all of the crop residue that we "waste" each year, but you have to have a pretty limited understanding of the role that crop residues play in maintaining the health of the soil to think that we can harvest it with no cost to the productivity of the land. The same goes for growing and harvesting algae from the ocean — another potential "solution." The sunrays that the algae would intercept are the primary source of energy for the ocean ecosystem. We won't capture them for free. We will be disrupting what's left of our ocean resource, which isn't much.

Biofuels are constrained by one fundamental biological function: photosynthesis. Photosynthesis is an incredibly inefficient method to capture light energy, convert it to chemical energy, and store it. The theoretical maximum conversion of light energy is 11%. That is, of the light energy that hits a plant, the absolute maximum that can be stored as chemical energy is 11%. In reality, most plants operate at 3 to 6% efficiency, and then for only part of the year. For example, the corn plants in a cornfield in our part of the world only cover 100% of the area of the field from early July until September.

I don't understand how our governments can continue to pour millions of dollars in subsidies into an industry that is so clearly a dead end. The only logical renewable energy development paths that we should be pursuing are wind, solar, and geothermal. Solar needs less than an order of magnitude improvement in efficiency and/or costs of production to be competitive with existing electrical power sources across most of the world. Most experts believe we will achieve that before 2020. Our future will be powered by electricity — electricity that isn't generated by burning fossil fuels. Transportation research should be focused on electricity-powered vehicles, not improving the efficiency of carbon-fuelled vehicles and adapting them to run on pure ethanol.

The good news is we won't hit a brick wall; there are enough fossil fuel reserves that we will be afforded a long tail of fossil fuel consumption overlap to allow the built infrastructure to be converted.

The price of renewable electrical energy will be coming down as the cost of fossil fuels rises. It should be a reasonably orderly conversion over the space of a century (notwithstanding the predicted impacts on our climate if we continue to add carbon to the atmosphere for the next century). It will be dragged on longer than it should, however, because powerful lobbies will succeed in convincing governments to invest tax dollars in their particular crazy scheme to create a "renewable" carbon energy source. It will be dragged on longer than it should by companies (and the people who invest in those companies) that continue to explore for oil and gas in sensitive ecosystems.

I know I'm just a farmer from the back concessions, but from where I'm sitting and according to the math I was taught in school, when it comes to growing crops for ethanol and biodiesel, those dogs just don't hunt.

CHAPTER 15

THE NAKED TRUTH

The bosom can ache beneath diamond brooches; and many a blithe heart dances under coarse wool.

— *Edwin Hubbel Chapin*

All farm animals are domesticated from prey animals. Because of this, they have different eyesight systems than humans do. Their eyes are on the sides of their heads instead of the front, giving them excellent peripheral vision but poor depth perception. They are also more acutely attuned to movement than humans. You can stop being a threat to an animal simply by stopping moving. One of the first steps when we're starting to halter train calves for the fair is to simply walk into the animal's pen and stop with some grain in your hand. Eventually their curiosity will overcome their fear, and they will approach — very cautiously at first, but trust will gradually be built up.

Silvia found out the opposite is also true. One beautiful autumn day, she had moved the sheep from one paddock to the next and was refilling their water trough. It was going to take a few minutes so she flaked out on the ground to soak up the sun, the breeze, and the pastoral setting. While she was lying down, an ewe came up to the trough to get a drink. The ewe mustn't have noticed Silvia laying there because when Silvia sat up, the ewe jumped straight up, did a 180-degree turn in the air, scampered away to a safe distance, and then stood there staring at Silvia. Silvia thought she was getting "the look." I think it was more likely that the ewe was looking at Silvia trying to figure out what she was. From an ewe's perspective, humans walk upright on two legs and can be seen approaching from a distance. They do not rise suddenly out of the ground.

Sheep have trouble recognizing anything that is a deviation from the ordinary. When we first started out with sheep we had less than 20. At shearing time, we first penned them up and then took each sheep from the pen, sheared it, and returned it to the pen. When you put the first couple of sheared sheep back in the pen, many of their flock mates won't recognize them. How do I know? Because the other sheep will be in a circle surrounding the sheared one looking at it and sometimes giving it a head butt or two. Once several are back in the pen, they tend to form their own group. Now that we're handling a lot more sheep, we have modified our cattle handling facilities slightly so that we can run the sheep through as well. The sheep come out of one pen, get sheared, and then get returned to a different pen.

Shearing is one of those jobs that you are glad when it's done and can't wait to have a shower. Not because you're itchy and scratchy as some people imagine because of their experience with factory wool garments, but because you are covered in lanolin. After shearing and handling every single one of our sheep, our pants are almost waterproof. Our hands are well moisturized and soft, but they smell like sheep. The inside of a fleece from our Romneys while it's still warm is incredibly soft. It's unfortunate that very few wool garments exhibit anything close to that softness.

Most people ask us how our wool can be so soft. They've never felt anything so soft and without the "itch" factor. The fleece of a sheep is generally shorn once a year. Over the course of the year, little bits of vegetable matter collect in the fleece — burrs from weeds and bits of hay and straw from their feed and bedding. In the industrial process, the fleece is first washed to remove dirt, sweat, and oils from the fleece. It's then treated with sulphuric acid and baked to cause the cellulosic plant material to carbonize. The carbonized plant material is crushed into a powder and falls out of the fleece. While the process is efficient for removing vegetable matter, it causes changes to the surface of the wool fibres. The overlapping scales that form the outside layer of the wool fibres are slightly lifted, which makes the wool scratchier.

It doesn't have to be this way. Many cottage woolgrowers have developed techniques to keep the fleeces sufficiently clean so that the acid bath can be skipped. Our wools are processed using a citrus-based soap and then a picker to remove most of the vegetable matter. You will be able to spot the odd bit of straw or hay if you look closely at our yarns, batts, and rovings, but the wool is softer and no sulphuric acid had to be produced, used, and disposed of. The industrial process changes the properties of wool in the name of efficiency.

The commercial wool industry has found a different solution to the problem: superwash wool. This wool goes through an additional process that coats the fibres with a man-made resin, smoothing the scaly surface. This imparts softness and machine washability. The scales on the surface of the wool fibres ratcheting against each other are the cause of wool garments shrinking and felting. There's one other secret of superwash wools. Most of them come from merino or merino-cross sheep. That's well known, as merinos produce some of the finest diameter fibres of any sheep breed. What's not well known, however, is that the closer to starving the sheep are, the finer the fleece that they produce.

A university classmate of mine, who is a large animal veterinarian, recently came back from a trip to Australia. She had toured a number of farms, including a large sheep station that specialized in producing fine wools. My classmate was appalled at the body condition score[84] of the sheep. The first question she asked me when I mentioned we'd added wool and yarns to our business was "Do you starve your sheep?" My answer was an unequivocal no. She then related her experience in Australia. The worst part of it is the animals have to be given the same level of nutrition day in and day out to maintain a constant thickness in the wool. We shear right around lambing in part because the extra nutrition demands on the ewe from the third trimester of gestation and the switch to starting to milk can cause a weakness or "break" in the fibre. The finest wools come from castrated male sheep (wethers) that are continuously fed a diet that just keeps them alive, not much more.

I've spent a lot of time in this book looking at the intersection between agriculture, food, the environment, and our health. Most of you are acutely aware of the fact that the body can absorb anything that you swallow. You are, essentially, what you eat. Do you know what the largest organ in the human body is? Skin. That's right, the membrane that covers your body weighs more than any of your other organs. The skin is also semipermeable. Compounds that touch the skin can be absorbed into the body. The palms of your hands, soles of your feet, your armpits, and your groin have the highest rate of absorption.

Do you put much thought into the fibres that you clothe yourself with? They spend 12 to 16 hours against your skin every day. An ever-

---

84. The body condition score is a technique for visually appraising the health and level of nutrition of all classes of four-legged animals. They are rated on a five-point scale where 1 is starving and 5 is obese. A well-managed herd or flock targets a 2.5 to 3.0 average body condition.

increasing percentage of clothes are made from some form of plastic fibre. They are treated with dyes, chemicals to resist stain, chemicals to create "permanent press," and fire retardants. In turn, each of the chemicals applied to clothing has been found toxic and replaced with newer chemistry that is currently deemed safer. One of the key ingredients in Scotchgard™ had a half-life in humans of 5.4 years. That means that if a given quantity of the compound entered your body today, half of it would still be there 5.4 years from now. The original permanent-press compounds were members of the urea-formaldehyde family of compounds.[85]

When it comes to fire retardants, the EPA has the following to say: "These studies have suggested potential concerns about liver toxicity, thyroid toxicity, developmental and reproductive toxicity, and developmental neurotoxicity. These findings raise particular concerns about the potential risks to children."[86] If all materials used to make clothing had flash-fire potential, I could understand a need for fire retardants. Applying fire retardants to clothing, however, only became necessary with the advent of man-made fibres being used in clothing — starting with rayon. Incidents of children quickly being engulfed in flames from a cigarette ash or a candle flame caused governments to mandate fire retardants in sleepwear. It was a case of our switch to a flawed synthetic fabric that caused the need for the application of these chemicals. With the increasing knowledge about the toxicity of fire retardants, children's sleepwear that meets stringent design criteria is exempted from being doused. The designs try to limit the possibility of the garments accidently being ignited.

Wool is a naturally fire-resistant, natural fibre. It's classed as self-extinguishing, meaning that it requires an external source of heat and flame to continue to burn. Once the flame is removed, wool stops burning. But who could sleep in wool pyjamas? They're itchy and scratchy and just plain uncomfortable. You now know that wool doesn't have to be. Our wool is so soft that at least one customer has knit a bra with it. (It's amazing what people are knitting these days. There was an entire issue of a knitting magazine dedicated to knitting lingerie.)

We sell both natural wool and dyed wool. A lot of people ask us about our dyes. Do we use natural dyes? The definition of a "natural"

---

85. The primary risk from these compounds is the "off-gassing" of formaldehyde, which can cause irritation of eyes, nose, and throat, and may cause cancer. The levels of formaldehyde released are highest when a garment is new.
86. U.S. Environmental Protection Agency, "Polybrominated."

dye is challenging. Many plant materials can be used to dye wool; however, you generally need to use a mordant along with the natural source materials to get bright and fast colours. Many mordants are salts of heavy metals. The primary mordant you can use that isn't a heavy metal changes the texture of the wool. We've been down many roads with our dyestuffs. We started with a line of weak acid dyes that were set with vinegar, and the dye solutions would completely exhaust — there was no dye material left in the waste water. We've tried several different techniques with natural materials and continue to experiment with them. The solution we have settled on for now is a line of dyes that uses no heavy metals in the dyes or in the manufacturing process. Will we change in the future? Yes, if something better comes along.

Let's put wool in the context of our other fabric choices. Cotton and silk are the two other major fibres. Conventional cotton covers about 3% of agricultural acres worldwide, but over 25% of all insecticides and 10% of all pesticides used by agriculture are used on cotton. Cotton is the most chemically dependent crop grown by conventional agriculture. The destruction of the Aral Sea was caused in part by diverting water for irrigating cotton fields.

Silk production generally doesn't involve the use of pesticides. The primary detraction from silk is the silk moths being boiled alive inside the cocoon to prevent them from harming the silk threads when they make their escape. Essentially, a silk cocoon is a single continuous fibre. To escape from the cocoon, the moth secretes an alkaline substance that dissolves a hole in the cocoon, which allows it to exit. That hole cuts the silk fibre into thousands of short lengths rather than one continuous fibre. I'm not overly concerned about the death of the moths. Their lives are shortened by at most ten days. The ones that emerge mate, lay eggs, and die.

Hemp and linen are also fabrics derived from plant material. Most linen is produced from conventionally grown flax. There is a resurgence of interest in hemp as a natural fibre source. Hemp's thick plant stand and tall stature lends itself to organic production. The biggest challenge with hemp fibre production is the lack of processing facilities in North America. Hemp production was banned almost a century ago. For the last decade, Canadian farmers have been able to grow hemp under a special license from the Canadian Food Inspection Agency. Most growers have focused on seed production for hemp foods simply because we have no fibre processing infrastructure. Raw hemp plants are very bulky and too costly to transport any significant distance. I grew hemp once. I filled out all the paperwork and had the inspections done to prove that the levels of THC were too low to give anyone a "buzz." Harvesting was the biggest challenge — putting hemp fibre through a combine while still green leads to wrapping and plugging problems (and sometimes fire). We put hemp in the "life is too short column."

After the natural fibres you have all the various synthetic fabrics. All are created from some form of plastic or a process that is very similar to plastic manufacture, and they come with all the associated environmental impacts and unknowns. There are a number of fibres that have appeared on the market recently claiming to be eco-friendly alternatives. They're made from bamboo, soybean, and corn. They are essentially rayon, which uses cellulosic material in the fibre-making process rather than petrochemicals. Most of the production methods involve a harsh chemical-based hydrolysis process followed by several rounds of bleaching. Some manufacturers have developed more environmentally friendly processes, but you need to do your research. Many companies' products aren't nearly as eco-friendly as they claim. Just because they start with a renewable wood or agricultural crop doesn't mean that the process to get from there to a soft fabric isn't polluting.

Even if you select a garment made from a natural fibre, you need to understand its history to be assured that you are wearing something sustainable. For example, there is one line of wool yarns available that wraps itself in feel-good-values marketing. The wool fibre, however, is sourced in Peru, transported to Scotland for spinning, sent to New Mexico for dying, and then transported to the company's corporate warehouse. How do I know this? Silvia bought a skein because she liked the colour and the ethic it portrayed. It wasn't until I read the fine print on the label that we realized how ridiculous the supply chain was.

It's possible to get clothing that has a small environmental footprint and isn't akin to wearing a toxic patch all day, but you need to educate yourself on the alternatives. There are a lot of unnecessary chemicals applied to your clothing in the name of convenience. It's the same as your food. You can choose to wear environmentally friendly clothing or you can choose to absorb toxins through your skin, pollute fields, support child labour, and pollute the developing world. It's your choice.

CONCLUSION

RE-RE-REGENERATION

There can be no purpose more inspiriting than to begin the age of restoration, reweaving the wondrous diversity of life that still surrounds us.

— *Edward O. Wilson*

One of the best Castor canadensis (Canadian beaver) live habitat recreations I have seen is in the Arizona-Sonoran Desert Museum. (It is better described as a historically accurate zoo of the flora and fauna that used to inhabit the region.) The Canadian beaver is indigenous to the Sonoran desert, which stretches from Arizona to California and Mexico, including the Baja California peninsula. Is beaver the first animal that jumps to mind when I say desert? It certainly caught me by surprise. The disconnect is created by the difference between what the Sonoran desert landscape was 150 years ago and what it is today.

Here's a reference to a valley in the Gila-San Simon area of Southeast Arizona: "Historical descriptions of the valley in the 1880s emphasize the ideal conditions for cattle. Bottomlands were covered with 65 cm tall grasses, open areas were dominated by grama and water was abundant on the valley floor."[87] Is that your image of the Sonoran Desert in Arizona and northern Mexico? It sure wasn't mine, but it explains how beavers could have survived and even thrived there. Historical precipitation records show that rainfall in the area hasn't changed much in the intervening century. What have changed are the demands agriculture has put on the ecosystem. Essentially, the desert was grazed to death. Without full vegetative cover, less rain soaks into the soil and what does soak into the ground evaporates quickly. Without water soaking into the landscape, the springs that once

---

87. McClaran, *Grassland*, 250.

fed the creeks and rivers dried up. Without vegetation, daytime temperatures are hotter, and the cycle continues.

The failure of the carbon cycle is a less obvious but devastating consequence of the failing conditions of the Sonoran desert. The grasslands once held substantial quantities of carbon in the vegetation and, more importantly, in the roots and soils. Those same soils today hold a small fraction of the carbon they did a century ago. The same can be said for the deserts of Libya, Ethiopia, Turkey, Iraq, Afghanistan, Central Asia, and the Atlantic coast of southern South America. All of these ecosystems have failed because humans overburdened them. Once the downward spiral started, humans' innate desire to survive simply worsened the problem. As yields and productivity fell, more land was cleared and grazed. As more land was grazed and cleared, less water was absorbed by the landscape and available to grow crops.

The good news is these collapses can be reversed. Agriculture can be used to reverse the devastating impacts of human activity on the planet. We can regenerate the planet while feeding everybody. That is my big, hairy, audacious goal (my BHAG). Agriculture can regenerate the planet while feeding everybody.

Allan Savory has proven it on a large scale. His TED talk video documents work completed using holistic planned grazing in the Sonoran desert in Mexico, in West Texas, and in central Africa. The restoration of these ecosystems is dramatic — the transformation of the landscape is almost too fantastic to believe. To do so, he simply changed the pattern of cattle grazing. He didn't plant any seeds. He didn't burn any diesel fuel. He didn't sprinkle any magic potions. That's part of the problem. There's nothing to patent and sell except wisdom that can be transferred in a several day workshop and a couple of workbooks. (I know this because I've been through the workshop and used the workbooks.) Oh, and the transformation takes almost a decade to complete.

While the transformation of the landscape has been incredible, it also includes the capture and permanent storage of significant quantities of carbon — a process that can continue indefinitely. Savory's Holistic Management can regenerate entire ecosystems as well as suck enough carbon out of the atmosphere to reverse climate change. Not only that, this revolution can also eliminate subtherapeutic use of antibiotics, reverse soil degradation (stop erosion and start building soils), improve water cycling and begin refilling the aquifers we have drained, restore the natural flow of rivers, virtually eliminate herbicide use, substantially reduce pesticide use, significantly reduce the energy embedded in our food, improve the humane treatment of animals, and restore the economies of the most impoverished regions of the world. Countries implementing these methods will not go into debt. On the contrary, their economies will improve and their societies will flourish.

Thank you John Ehrenfeld for your revised definition of sustainability:

Sustainability is the possibility that humans and other life will flourish on Earth forever.[88]

Climate disruption is reversible and agriculture is the single largest and cost-effective tool to accomplish it using technology that already exists. We don't need bioengineers to create some magical plant or animal. (In my not-so-humble opinion, we would all see the solution a lot faster if bioengineers stopped offering false hope.) We can start today. Some of us already have. We just need more people to see the solution. These ideas need to spread.

This transformation is going to require some sacrifices. Vegetarians are going to have to start eating grass-fed beef, lamb, and goat instead of ecosystem-destroying soybeans. All of us will need to stop eating processed food, since that's where the real carbon emissions problem in the food supply is found. Society will have to accept sewage sludge spreading as a form of true recycling. We will have to stop putting anything that isn't biodegradable down our drains or toilets and become comfortable eating pork and chicken that have up-cycled food waste. We will have to consistently separate our reprocessing streams. (There can be nothing called "garbage" in a flourishing, sustainable society). Last, we will have to postpone retirement because we will all be healthier and live longer. Think you can handle that?

Here is a list of simple rules that should guide your choices towards a planet-regenerating diet:

1. Establish a relationship with your farmer(s).

There is no label that can substitute for a relationship. You need to know where your food is being produced and under what conditions. You need to know that the farmer has similar values to you with respect to quality, the environment, and humane treatment.

---

88. See note 7.

Whether you gather that information virtually or in person is your choice. For example, I will likely never meet my banana farmer face to face (and no, I haven't found a good source for bananas that meet all my criteria — as I said, we're a work in progress). However, you can be assured that slick TV commercials and magazine advertisements are not the places where I would start.

2. Eat more perennials.

We need to switch our diets to largely perennial crops to ensure a flourishing future. I don't mean large, sterile almond orchards with the bees brought in by the tractor-trailer load at pollination time; I mean perennials that are managed as part of a well-functioning ecosystem. Ecosystems are diverse. For example, to sustain pollinators, there needs to be a food source (flowers) throughout the year. To maximize water infiltration, the soil should be covered with growing plants throughout the year. To maximize carbon sequestration, a diverse ecosystem of plants with varying root depths and growth habits is needed.

The majority of your current diet is likely derived from annual crops. All your cereals, sugar, meat, and vegetables are annual crops. Most chicken and pig feed is derived from annual crops. Most beef animals are finished on annual crops. The diet of most dairy cows contains a healthy portion of annual crops. On the other hand, most nuts (though not peanuts) are perennials. Most food items that we consider fruit (though not tomatoes or pumpkins) are perennials: apples, peaches, pears, plums, cherries, raspberries, bananas, mangoes, figs, grapes, pineapple, etc. The bottom line is to eat more perennials and animals that eat perennials.

3. Pigs and chickens are not vegetarians; so don't be alarmed if they are being fed food waste or by-products (brewers' grains, blood meal, feather meal, cottonseed meal).

In a flourishing future, pigs and chickens are tasked with converting food and plant waste into human-edible food. Their diet should be diverse and viewed as a chance to up-cycle waste and by-products that otherwise would be composted or landfilled. Anyone trying to sell you "vegetarian-diet" chicken, eggs, or pork isn't doing what is in the best interest of the future or worthy of your patronage.

4. Only buy rotationally grazed, 100% grass-fed beef and lamb.

This is very simple. Purely grass-fed beef and lamb are healthier for you, the environment, and the animals themselves. Properly managed rotational grazing is one of the few tools that can be deployed immediately on a large scale to sequester carbon and rebuild degraded soils.

5. Don't flush anything down your toilet that you don't want spread on the ground that grows your food.

The nutrients in your food can be cycled indefinitely through the food system. We can solve a number of problems by using sewage as fertilizer. One of the biggest challenges currently is all the other stuff that people flush into the sewers. The sewer system isn't a magician. Nothing that goes into it disappears; it is just relocated.

6. Cook your own meals.

The single biggest change you can make to improve your diet and your impact on the future is to start cooking your own meals. Most food processing is a waste of energy and most processed foods have far more salt and sugar than what is healthy — and they are the additives that you should be most worried about. When you cook, you are in control of your diet and the future of food production and the planet.

To help you start the conversation in your search for a flourishing future, I've compiled a list of 18 questions for your discussions with your potential farmer(s):

1. Are you certified organic?

2. Do you rotationally graze?

3. Do you feed your cattle and sheep any grain?

4. Are your chickens and pigs fed waste products?

5. How do you use antibiotics?

6. How much time do your animals spend outside?

7. Do you castrate, dock tails, trim beaks, or make any other surgical modifications to animals?

8. What are you doing to sequester carbon?

9. How often do you till the soil?

10. What are you doing to eliminate erosion?

11. What are the sources of fertility on your farm?

12. Do you raise any GMO crops or animals?

13. What are you doing to protect groundwater and surface water?

14. What are you doing to maximize your harvest of rainwater? (And if you irrigate, how do you know your water source for irrigation is sustainable?)

15. What is the source of labour for your farm?

16. What are you doing to protect biodiversity (plants, animals, and wildlife)?

17. How much of the energy you use is from nonrenewable sources?

18. Can I have a tour?

Many of these questions don't have right or wrong answers. This isn't a checkbox exercise. Your goal is to learn about the farmer's world view and values and how they manifest themselves in the practices on the farm. Recognize that what I'm describing is a long-term shift in thinking and an exercise that can't be accomplished overnight. I've described a transformation on our farm that has taken over 15 years and still continues to change. We're not done. If you want perfection, don't shop at our farm because we don't have it completely right. In fact, I'm not sure what perfection is, but I do know that perfection isn't found in a particular product or in a single business. Perfection will be found in the systems surrounding the products. I can't claim to be perfectly sustainable until every business I interact with is also sustainable.

As you seek out a flourishing future, don't try to take shortcuts by using labels as your guides. If you want to put a label on my world view, you'll need several. I am a husband. I am a father. I am Christian. I am an environmentalist. I am a capitalist. I am a libertarian. I am an economist. I am a farmer. I am a businessman. When you role all of that together, you get someone who has faith in the good of people and believes we can come together to solve problems that are much larger than ourselves without creating massive government bureaucracies. Individual freedom and choice will prevail when people are given full information about the consequences of their actions.

I challenge you to ask questions and learn. You want to be able to make an informed decision. In the end, a flourishing food system and society is your choice to make.

You can choose to find time to cook more of your own meals from real food, or you may be forced to find time to suffer from any one of a number of chronic diseases.

You can choose meat produced without subtherapeutic antibiotics, or you may have to attend the funeral of a loved one that died from an antibiotic-resistant infection.

You can choose food raised by a farmer focused on building soil, or you can watch documentaries with your grandchildren on how the soil was destroyed.

You can choose to actively participate in and shape the system that delivers your food, or you can eat whatever the system decides to serve you.

You can choose to commit to the person that produces your food and form a mutually beneficial relationship, or you can consume food produced by the lowest-cost producer.

You can choose to eat fossil food, or you can choose regenerative food.

You can choose organic food, or you choose continued natural gas exploration and fracking.

You can choose to improve your health and the health of the environment, or you can choose to improve the health of multinational corporations' bottom lines.

You can choose to fill a sandbag, or you can wait until the government does it for you (but don't hold your breath).

You can choose your future.

You are choosing your future.

You are choosing our future.

Please choose a flourishing future.

APPENDIX

Here are my answers to the questions:

1. Are you certified organic?

Our land is certified organic and any crop that we produce is organic; however, none of our meat is certified organic. We don't use the term "organically raised" because it is meaningless. We ended up here by default. When we first started raising sheep and cattle, they were fed with hay from land that we were transitioning to organic, so they couldn't be certified the first couple of years. With the pigs, we started out buying weaned piglets from other farmers, but we couldn't find anyone selling certified organic weaner pigs; therefore, we couldn't certify them either. Our abattoir isn't certified organic, but we could have worked around that with some extra paperwork. Thus, we started out the meat CSA program with meats that couldn't be certified organic. Once we were in a position where we could certify them, we asked our customers how important various parts of our methods and philosophies were. Certifying our meat wasn't even close to the top of their list. So rather than spending time on certifying our production, we focused on raising the best grass-fed and pastured-raised meats we could.

2. Do you rotationally graze?

Yes. We follow the Planned Holistic Grazing approach advocated by Allan Savory. We are averaging over 200 days of grazing annually and hope to continue to increase that number.

3. Do you feed your cattle and sheep any grain?

No.

4. Are your chickens and pigs fed waste products?

Currently, the only waste they are fed comes from our kitchen. We are working on sourcing brewers' grain from a craft brewery to form a portion of our pig and chicken diets, but the logistics are proving to be challenging. The majority of their diet is a standard non-GMO grain ration.

5.   How do you use antibiotics?

We use no subtherapeutic doses of antibiotics. We do not routinely give antibiotics. The only time we use antibiotics is when an animal has an infection that needs to be treated.

6.   How much time do your animals spend outside?

They spend as much time as they want outside. All our animals have access to the outdoors for the majority of their lives. There are some minor exceptions, such as chicks in a brooder, newborn piglets in winter, and sheep for a few days before shearing to ensure the fleeces are as dry as possible. We use modified Joel Salatin practices for our broilers and hens.[89]

7.   Do you castrate, dock tails, trim beaks, or make any other surgical modifications?

Mostly no. We do not castrate, trim beaks, or dock pig tails. We dehorn our commercial cattle if they need it — we are crossing in hornless (polled) genetics to eliminate this procedure. We also remove lamb's tails by elastration — a rubber band is put around the tail to cut off blood flow, the tail tissue dies, and then it falls off. We do this to reduce the amount of faecal matter and urine that collects in the wool on a sheep's rear end. During our warm, humid summers, the trapped faecal matter can become a cause of several nasty infections and conditions. Even with taking this precaution we have had a few incidents. From our point of view, the minor distress caused by removing the tail is well worth preventing the other conditions.

8.   What are you doing to sequester carbon?

We are 100% perennial pasture and hay. None of our soils have been tilled since 2011 and some were last tilled in 2005. We have planted innumerable trees and are building biomass in our forested land by selective harvesting.

---

89. "Joel Salatin practices" refers to housing chickens in structures that are moved across pastures to ensure the chickens always have fresh green forages available and to prevent them from denuding any one spot.

9. How often do you till the soil?

Our plan is to never till soil again.

10. What are you doing to eliminate erosion?

See answer to question #8.

11. What are the sources of fertility on your farm?

We are primarily recycling the fertility that exists within our soils and working to increase the biological activity and depth from which plants can draw nutrients. We are importing some fertility through the feedstuffs we bring onto the farm to feed our pigs and chickens. We have experimented with some organic soil amendments, but we have not used anything routinely.

12. Do you raise any GMO crops or animals?

No.

13. What are you doing to protect groundwater and surface water?

See answer to question #8. As well, we have improved the heads of our wells to eliminate the possibility of surface-water infiltration. There is a wide buffer of forested area between the land that we farm and the river that flows through it.

14. What are you doing to maximize your harvest of rainwater? (And if you irrigate, how do you know your water source for irrigation is sustainable?

See answer to question #8. We do not irrigate.

15. What is the source of labour for your farm?

Our family members provide all of our labour. Each of the children has a financial interest in at least one of our enterprises.

16. What are you doing to protect biodiversity (plants, animals, and wildlife)?

In addition to our diverse perennial pastures, we raise a number of endangered breeds, including White Park cattle, Romney sheep, Berkshire pigs, and several rare breeds of poultry and fowl. We are re-establishing fencerows across the farm. Within the tree lines, there are a number of deciduous trees with edible nuts and fruit for the birds and wildlife. We have participated in university research into the nesting habits of two endangered species of birds — bobolink and eastern meadowlark — both of which are found on our farm.

17. How much of the energy you use is from nonrenewable sources?

All of our heating is provided by wood sourced primarily from our own bush. We still use petroleum fuel to harvest hay, but are working to reduce the amount of hay we need to feed by increasing the length of our grazing. Our electricity is sourced directly from the grid, but we have plans to install some solar in the future.

18. Can I have a tour?

Yes, but please call to arrange a mutually convenient time or watch our website for an opportunity to join an already-scheduled event at the farm.

REFERENCES

Alderman, Michael H. "Colin Johnston – A Celebration: Salt, Blood Pressure, and Human Health." Hypertension: Journal of the American Heart Association 36 (November 2000): 890–893. doi: 10.1161/01.HYP.36.5.890.

Badgley, Catherine, Jeremy Moghtader, Eileen Quintero, Emily Zakem, M. Jahi Chappell, Katia Avilés-Vázquez, Andrea Samulon, and Ivette Perfecto. "Organic agriculture and the global food supply." Renewable Agriculture and Food Systems 22, no. 2 (June 2007): 86–108. doi: 10.1017/S1742170507001640.

Chang, Feng-chih, Matt F. Simcik, and Paul D. Capel. "Occurrence and fate of the herbicide glyphosate and its degradate aminomethylphosphonic acid in the atmosphere." *Environmental Toxicology and Chemistry* 30, no. 3 (March 2011): 548–555. doi: 10.1002/etc.431.

Colborn, Theo. "Neurodevelopment and Endocrine Disruption." *Environ Health Perspect* 112, no. 9. (June 2004): 944–949. doi: 10.1289/ehp. 6601.

Diamon, Jared. *Guns, Germs and Steel: The Fates of Human Societies.* New York: W. W. Norton & Company, 1997, 1999.

Edwards, Clive A. "The concept of integrate systems in lower input/sustainable agriculture." *American Journal of Alternative Agriculture* 2, no. 4 (Fall 1987): 148–152. doi: 10.1017/S0889189300009255.

Ehrenfeld, John R. *Sustainability by Design: A Subversive Strategy for Transforming Our Consumer Culture.* New Haven: Yale University Press, 2008.

Ernst, N., M. Fisher, W. Smith, T. Gordon, B. M. Rifkind, J. A. Little, M. A. Mishkel, and O. D. Williams. "The association of plasma high-density lipoprotein cholesterol with dietary intake and alcohol consumption. The Lipid Research Clinics Prevalence Study." *Circulation: Journal of the American Heart Association* 62 (July 1980): 41–52.

Fleming, Sir Alexander. "Nobel Lecture: Penicillin." *Nobel Lectures, Physiology or Medicine 1942–1962*. Amsterdam: Elsevier Publishing Company, 1964.

Halberg, Niels, H. F. Alroe, M. T. Knudsen, and E. S. Kristensen, eds. *Global Development of Organic Agriculture: Challenges and Prospects*. Cambridge: CABI Publishing, 2006.

Hinderer, Julie Mida, Michael W. Murray, and Trilby Becker. "Feast and Famine in the Great Lakes: How Nutrients and Invasive Species Interact to Overwhelm the Coasts and Starve Offshore Waters." National Wildlife Federation, 2011.

Holmgran, David, and Bill Morrison. Permaculture One: A Perennial Agricultural System for Human Settlements. Tagari Publications, 1978.

Howard, Barbara V, and Judith Wylie-Rosett. "Sugar and Cardiovascular Disease: A Statement for Healthcare Professionals From the Committee on Nutrition of the Council on Nutrition, Physical Activity, and Metabolism of the American Heart Association." *Circulation: Journal of the American Heart Association* 106 (July 23, 2002): 523–527. doi: 10.1161/01.CIR.0000019552.77778.04.

Kong, Angela Y. Y., Johan Six, Dennis C. Bryant, R. Ford Denison, and Chris van Kessel. "The Relationship between Carbon Input, Aggregation, and Soil Organic Carbon Stabilization in Sustainable Cropping Systems." *Soil Science Society of America Journal* 69, no. 4 (2005):1078–1085.

Leopold Centre for Sustainable Agriculture. "The Long-Term Agroecological Research (LTAR) Experiment." Iowa State University, revised August 2012. http://www.leopold.iastate.edu/sites/default/files/pubs-and-papers/2012-08-long-term-agroecological-research-ltar-experiment.pdf.

Liu, S., W. C. Willett, M. J. Stampfer, F. B. Hu, M. Franz, L. Sampson, C. H. Hennekens, and J. E. Manson. "A prospective study of dietary glycemic load, carbohydrate intake, and risk of coronary heart disease in US women." *The American Journal of Clinical Nutrition* 71, no. 6 (2000): 1455–1461.

MacRae, Rod, Ralph Martin, Anne Macey, Paddy Doherty, Janine Gibson, and Robert Beauchemin. "Does the adoption of organic food and farming systems solve multiple policy problems? A review of the existing literature." Published online by The Organic Agriculture Centre of Canada. January 30, 2004. http://www.organicagcentre.ca/Docs/Paper_Benefits_Version2_rm.pdf.

McClaran, Mitchel P., and Thomas R. Van Devender, eds. *The Desert Grassland*. Tuscan: University of Arizona Press, 1997. Malthus, Thomas. *An Essay on the Principle of Population*. Reprint of the 1798 edition, Project Gutenberg, 2003. http://www.gutenberg.org/ebooks/4239.

Martindale, Wayne, and James J. Vorst. "The Rothamsted Long Term Agricultural Experiments." MPC Research, 2003. http://www.foodinnovation.org.uk/download/files/Corporate_Social_Responsibility_research/Dr_Wayne_Martindale_OECDfellowship1004_(1).pdf

Millennium Ecosystem Assessment. *Ecosystems and Human Well-being: Synthesis Health*. Geneva: World Health Organization, 2005.

"Nutrient intake and its association with high-density lipoprotein and low-density lipoprotein cholesterol in selected US and USSR subpopulations. The US-USSR Steering Committee for Problem Area I: The pathogenesis of atherosclerosis." American Journal of Clinical Nutrition 39, no. 6 (June 1, 1984): 942–952.

O'Brien, Thomas F. "Emergence, Spread, and Environmental Effect of Antimicrobial Resistance: How Use of an Antimicrobial Anywhere Can Increase Resistance to Any Antimicrobial Anywhere Else." *Clinical Infectious Disease* 34, No S3 (2002): S78–S84. doi: 10.1086/340244.

Olson, Rob. "A program called Alternative Land Use Services (ALUS) might be our best hope for restoring duck production across prairie Canada." *Delta Waterfowl Magazine*, Fall 2009.

Rodale Institute. *The Farming Systems Trial: Celebrating 30 years.* 2011. Available at http://rodaleinstitute.org/our-work/farming-systems-trial/ farming-systems-trial-30-year-report/.

Savory, Allan. "How to fight desertification and reverse climate change." Filmed February 2013. TED video, 22:20. Posted March 2013. http://www.ted.com/talks/allan_savory_how_to_green_the_world_s_ deserts_and_reverse_climate_change.html

Savory, Allan, and Jody Butterfield. *Holistic Management: A New Framework for Decision Making*, 2nd ed. Washington: Island Press, 1999.

Snell, Chelsea, Aude Bernheim, Jean-Baptiste Bergé, Marcel Kuntz, Gérard Pascal, Alain Paris, Agnès E. Ricroch. "Assessment of the health impact of GM plant diets in long-term and multigenerational animal feeding trials: A literature review." *Food and Chemical Toxicology* 50, nos. 3–4 (March–April 2012): 1134–1148. doi: 10.1016/j.fct.2011.11.048.

Teasdale, J.R., L.O. Brandsæter, A. Calegari, and F. Skora Neto. "Cover Crops and Weed Management." *Non Chemical Weed Management Principles, Concepts and Technology*. Edited by M.K Upadyaya and R.E. Blackshae. Wallingford, UK: CABI, 49–64.

The World Economic Forum Water Initiative. *Water Security: The water-food-energy-climate nexus.* Edited by Dominic Waughray. Washington: Island Press, 2011.

United Nations. "Our Common Future: Report of the World Commission on the Environment and Development." Report transmitted to the General Assembly as an annex to document A/42/427. 1987.

United Nations Special Rapporteur on the Right to Food, and Oliver De Schutter. "Agroecology and the Right to Food." Report presented at the 16th Session of the United Nations Human Rights Council [A/HRC/16/49] on March 8, 2011.

U.S. Department of Health and Human Services. "FDA Docket No. FDA 2005-P-0007: Petition Denial." Written by Leslie Klux on November 7, 2011, to Andrew Maguire, Vice President, Environmental Health, Environmental Defense. http://www.regulations.gov/#!documentDetail; D=FDA-2005-P-0007-0007.

U.S. Environmental Protection Agency. "An Exposure Assessment of Polybrominated Diphenyl Ethers (External Review Draft)." U.S. Protection Agency, Washington, DC, EPA/600/R–08/086F.

ACKNOWLEDGEMENTS

First off, I'd like to thank all the various creatures that found their way into this book — especially Kasne, Lilly, and Cauliflower. Without their antics and stubborn refusal to conform to stereotypes, this book and my life would be a lot less entertaining.

My editors — Christa Bedwin, Sheila Wawanash, and Meghan Behse — you each served a specific purpose at the time you entered the life of this manuscript. Christa, your enthusiasm gave me a boost. Sheila, your firm but pragmatic editing forced me to really think about what I was trying to say and untangle both my thoughts and words. Meghan, your efficiency and thoughtful editing allowed me to finally put a bow on this book and call it complete.

To our loyal CSA members and newsletter subscribers, I wouldn't have embarked on this literary journey without your support and encouragement. Several of you even managed to make it through the first draft of the manuscript — I thank you for your endurance and diplomacy.

Alec Mills, I'm not sure where I'd be today without our friendship and partnership. The first step in this journey can be traced back to a conversation you and I had while looking at a field of white beans on a soil and crop tour. That was the day I started to consider organic a serious option.

My sister Sandra — you proved to be an excellent foil as I worked through many of the topics in this book and life in general. It's a good thing I don't have to pay for long-distance calls by the minute.

To my family — Silvia, Connor, Cavan, Harrison, Abigayle, and Rebekah — you've paid the true cost of this self-indulgent exercise through time I didn't spend with you so I could write and through chores you did for me when "the words were finally flowing." Thank you.

HARRY STODDART

Harry Stoddart holds a Bachelor of Science in Agriculture and a Master of Science in Agricultural Economics. He has completed agricultural policy analysis for the federal government and almost every provincial government in Canada. Harry is an associate at the George Morris Centre, Canada's foremost agri-food industry think tank, and he sits on the advisory board and is a faculty member of the Sustainable Agriculture Program at Fleming College, Ontario.

The Stoddart Family Farm pioneered the use of the Community Supported Agriculture (CSA) style of selling meat in Canada and now uses CSA share programs for meat, eggs, and wool. Harry's farm has the largest 100% grass-fed cattle herd and sheep flock grazing on certified organic pastures in Ontario. His customers tell him he has the best meat going.

Iguana Books

iguanabooks.com

If you enjoyed *Real Dirt: An Ex-Industrial's Guide to Sustainable Eating*...
Look for other books coming soon from Iguana Books! Subscribe to our blog for updates as they happen.

iguanabooks.com/blog/

You can also learn more about Harry Stoddart and his upcoming work on his blog.

http://stoddart.ca/category/day-in-the-life/

If you're a writer ...
Iguana Books is always looking for great new writers, in every genre. We produce primarily ebooks but, as you can see, we do the occasional print book as well. Visit us at iguanabooks.com to see what Iguana Books has to offer both emerging and established authors.

iguanabooks.com/publishing-with-iguana/

If you're looking for another good book ...
All Iguana Books books are available on our website. We pride ourselves on making sure that every Iguana book is a great read.

iguanabooks.com/bookstore/

Visit our bookstore today and support your favourite author.

IGUANA

CPSIA information can be obtained at www.ICGtesting.com
Printed in the USA
LVOW08s1053171013

357241LV00002B/21/P